APPRENTICES
of
WONDER

APPRENTICES
of
WONDER
Inside the Neural Network Revolution

William F. Allman

BANTAM BOOKS
NEW YORK • TORONTO • LONDON • SYDNEY • AUCKLAND

APPRENTICES OF WONDER
A Bantam Book / October 1989

Library of Congress Cataloging-in-Publication Data
Allman, William F.
  Apprentices of wonder : inside the neural network revolution /
William F. Allman.
    p.   cm.
Bibliography: p.
Includes index.
ISBN 0-553-05389-2
1. Neural circuitry.   2. Neural computers.   I. Title.
QP363.3.A45   1989
612.8′2—dc20   89-14940
                                                         CIP

Published simultaneously in the United States and Canada

PRINTED IN THE UNITED STATES OF AMERICA

B   0 9 8 7 6 5 4 3 2 1

For Patricia

The brain is an enchanted loom, where millions of flashing shuttles weave a dissolving pattern, always a meaningful pattern though never an abiding one.

—Sir Charles Scott Sherrington

My brain! That's my second favorite organ.

—Woody Allen
Sleeper

Contents

From Neuron to Psyche

The Brain—is wider than the Sky—
For—put them side by side—
The one the other will contain
With ease—and You—beside.
                    —Emily Dickinson

"NAAAMAAAMEEEMOONAAAAAAEEEEEEEAHHHHHH."
A thin, mechanical voice is wailing from the tape player cradled in Terry Sejnowski's lap. Sejnowski, a biophysicist at Johns Hopkins University, is trying to make his face match the professorial look of his red pin-striped shirt and tie, but his young features keep rebelling, slipping into a grin not unlike that of a father watching his child's first performance on the violin.

The voice from the tape recorder belongs to NETalk, a machine Sejnowski created that is teaching itself to read aloud. *"AAAH-NEEEE-OOOOEEEEE-CH-CH-CH-EEEEMIIIII."* The voice, high-pitched and monotonous, continues its bizarre, nonstop chant. A crazed mantra of coos and clicks fills the room with a noise that somebody might make if he were, as the joke used to go, "spending the first day with a new mouth."

"What we're hearing is the first training session," Sejnowski shouts over the noise. "It's reading the text, but it doesn't have any idea what it's doing. So it's basically just producing random sounds."

NETalk is not a typical computing machine. Unlike the serial digital computers that have become so much a part of home, business, and academic life, NETalk is modeled after a computer that comes from life itself—the computer in your brain. NETalk is made with artificial neurons, connected to one another in a vast network. When NETalk "thinks," the neurons "talk" to one another, propagating signals throughout the web of connections.

"EEEEEOOOOONEEEMAAAAAAOOOOOOOOOOOO."
The eerie cacophony from Sejnowski's tape recorder continues for

another minute or so. Then slowly, gradually, NETalk's wail begins to change, taking on an order of sorts. The voice still has a high-pitched, continuous tone, but it's no longer monotonous. The long vowel sounds are regularly broken into smaller bits by consonants. It's a sound that you might hear in a nursery.

"MAMAMAGAGAMAAMTATATMAMAMAMA." "Now it's going through a stage where it's babbling," says Sejnowski. "The first thing it discovers is the distinction between vowels and consonants. But it doesn't know which is which, so it just puts in *any* vowel or consonant. It babbles."

NETalk isn't designed like an ordinary computer; nor is it programmed like one. To train NETalk to read aloud, Sejnowski had given his machine a thousand-word transcription of a child's conversation to practice on. The machine was reading the text over and over, experimenting with different ways of matching the written text to the sound of spoken words. If it got a syllable right, NETalk would remember that. If it was wrong, NETalk would adjust the connections between its artificial neurons, trying new combinations to make a better fit.

A few more minutes of babbling pass, then suddenly NETalk's steady stream of chatter stops. But it's silent for only a second. NETalk quickly sputters sounds again, stops again, and restarts. *"Mop . . . Chi-Ah-Nee . . . Eee . . . Eee . . . Nib-an-pan-toe-nee."*

"Something really strange happens here," says Sejnowski. "Hear the difference? Now it's discovered *spaces,* the distinction between words. So it speaks in bursts of sounds, pseudowords." NETalk rambles on, talking nonsense. Its voice is still incoherent, but now the rhythm is somewhat familiar: short and long bursts of vowels packed inside consonants. It's not English, but it sounds something like it, a crude version of the nonsense poem "The Jabberwocky."

Sejnowski stops the tape. NETalk was a good student. Learning more and more with each pass through the training text, the voice evolved from a wailing banshee to a mechanical Lewis Carroll. "What you just heard was the result of only about an hour and a half of computer time," says Sejnowski. "What you'll hear next is what happened after we left NETalk on overnight. We just let it read the same text over and over for ten hours."

Sejnowski's father face has won completely now, and a broad smile beams as he turns on the tape deck. From the box spills

NETalk's voice, no longer a monotonous wail or mangled string of gibberish. The timbre is still somewhat mechanical, the inflection still a little out of kilter but the diction is definitely a few steps into the human side of the world: *"I walk home with some friends from school,"* says the machine. *"I like to go to my grandmother's house. Because she gives us"*—NETalk pauses a split second, struggling—*"candy."*

How does the brain work? This book was inspired by that seemingly simple question, and while I expected the answer to be complicated, I soon found that the question was also more complex than I had imagined. The brain is a monstrous, beautiful mess. Its billions of nerve cells—called neurons—lie in a tangled web that displays cognitive powers far exceeding any of the silicon machines we have built to mimic it.

This complexity extends the question of how the brain works beyond merely asking for an explanation of the biological processes of nerve action or a psychological description of the various aspects of the mind. The heart of the question lies in the connection between biology and psychology: How do the actions of the brain's billions of neurons produce the rich tapestry of the mind?

That is a difficult question indeed, and so it is not surprising that for the past quarter century few scientists studying the brain and the mind have dared to ask it. Instead, most researchers have concentrated on studying either the biology of the brain or the behavior of the mind, without attempting the enormous task of figuring out how one produces the other.

Recently, however, there has been a new effort to join the various pieces of brain and mind research into one science. Spawned by the enormous strides made in neurobiology and computer science over the past several years and fueled by a deep frustration with the conventional approaches to understanding how we think, a rapidly growing group of researchers has taken a new tack in trying to understand the machinery of the mind. In this new scientific endeavor—some of the more passionate researchers are calling it a *revolution* that rivals the emergence of relativity and quantum mechanics in physics at the beginning of this century—scientists are creating a new model of the mind, one that attempts to understand how the mind works by examining how the brain works.

Over the years researchers have discovered an abundance of

details about the brain and mind but, they are just now beginning to ask the fundamental question of how these parts interact as a unified system. Their efforts give nonscientists, perhaps a little jaded by the fantastic scientific achievements of the twentieth century, an ideal opportunity to observe how scientists behave when they aren't quite sure how to proceed.

This new endeavor in brain and mind science is still in its infancy, but its early findings are exciting because they suggest the new model of how we think may turn traditional scientific approaches upside down in every area of research involving the brain and the mind. From neuroscience, psychology, and computer science to linguistics, philosophy, and the social sciences, this new model of the mind may redefine not only the answers that are sought but also the questions that are asked. Along the way, it may also upend the way we think about ourselves.

As with most areas of science, there are more people involved in this new approach than can be accommodated between the covers of a book. I've focused on a small group of researchers, chosen as much for their eloquence and character as for their scientific contributions. These researchers should be seen as a representative slice, a profile of the scientists who are constructing the new model of the mind and creating a new way of thinking about the way we think. They are forging a new language with which we can ask questions about the mind and brain, and their work gives us a glimpse at the process of scientific revolution.

A book spanning this wide range of subjects is a group enterprise, and I thank all the scientists who gave many hours patiently explaining their research and reading parts of the manuscript. I also wish to thank Kristine Dahl, Peter Guzzardi, and Leslie Meredith for editorial help in bringing the book to fruition; Paul Allman, Terrence Monmaney, and Russ Rymer for helping to shape the overall concepts; and Claudela de Valdenebro for her artwork. Special thanks to Terry Sejnowski, who served as my guide through the tangled web of brain science, and John Tierney, without whose encouragement this book would not have been possible.

The answers to the fundamental questions about the brain and mind are yet to come. But most people who enjoy reading about science—like most scientists—know that often the hardest part of answering a question is knowing the right one to ask. This book is for them.

CHAPTER ONE

MAKING UP A MIND

The New Connectionist Revolution

MIND, n. —A mysterious form of matter secreted by the brain. Its chief activity consists in the endeavor to ascertain its own nature, the futility of the attempt being due to the fact that it has nothing but itself to know itself with.
—Ambrose Bierce, *The Devil's Dictionary*

The mind has been trying to understand itself for centuries. The dense lump of tangled nerve tissue, chemicals, and electricity has had far more success looking out, however, than looking in. It has divined all but the first instant of the history of the universe, unraveled the genetic machinery of life, and traced the origins of its own existence back millions of years to the first hominids. But sealed inside its thick shell of bone, our brain tissue tells itself very little about the processes it uses to create symphonies, learn the polka, or laugh at *I Love Lucy*.

Like the fabled blind men investigating an elephant, scores of philosophers, psychologists, neuroscientists, and other cognitive researchers have grappled with various parts of our mindworks, hoping to reveal its secret structures. They have been recently joined by computer scientists trying to create artificial intelligence by mimicking the mind on machines.

There have been some successes. Scientists have a general idea about the workings of the various nerve cells—called neurons—that make up the brain. They have traced some of the electrical conduits that carry information from our sensory organs to our thinking centers and located general areas of the brain where certain types of information processing—language, for example—take place. They've also gleaned some sense of what thinking is, at least as the human mind does it. They have a rough idea of what the mind does well, what it does not do so well, and the various ways it can malfunction.

But when it comes to explaining just how those neurons, chemicals, and electrical conduits produce the various phenomena of the mind, nobody really has the slightest idea where to begin. In a way, it boils down to a simple question: *What makes a bunch of neurons so smart?*

For years, scientists trying to answer that question have responded by splitting it in two. Neuroscientists, for example, study the "bunch of neurons" part of the question. Psychologists, linguists, computer scientists, and philosophers take on the task of describing what "so smart" means. In other words, some scientists study the brain's biology; others, the mind's behavior. In dividing the question, however, scientists left out a crucial part—explaining *what makes* it all happen.

This approach is a little like trying to understand how a television works by studying a wiring diagram that shows its transistors, diodes, and other pieces of electronic circuitry and then watching various television programs. The information wouldn't tell you anything about how the electronic circuitry makes the programs appear on the screen.

The brain, of course, is far more complicated than a television. In fact, the most significant result of the last three decades of research in cognitive science has been failure—and the realization that the mind is far more complex than a souped-up version of a pump, clock, telephone exchange, or any of the other machines scientists have modeled it after over the years.

That applies to the newest aspirant to be the model of the mind, the digital computer. Research in artificial intelligence—called AI for short—has produced some success stories. AI researchers have designed "expert" systems used in fields such as medicine and mining, and their chess-playing programs are ranked among the best players in the world.

But the early hope that computer science would unlock the secrets of the mind has not been realized. There is now a growing suspicion about the fundamental assumption, which has guided cognitive science research for decades, that the mind processes information in essentially the same way as a serial digital computer. Many researchers are beginning to wonder whether studying the mind by comparing it to a digital computer is like trying to find out how an airplane works by studying a helium balloon.

Not that computers aren't smart; in fact, computers can do some cognitive feats that would put the brain to shame. But our brains can sometimes make the most powerful computers look like Bozos. For example, we can recognize our mothers right after they've had a haircut. We can understand the speech of a Boston Brahmin or a tugboat captain from Baton Rouge. We can recognize something as a chair, whether it's a Chippendale, a beanbag, or a throne. Wee con undrestin wrds efen wen theh ar missspld. Or fill in the blanks when l t rs are missing. Many of us know immediately what the figure below represents.

And if we don't, when we are told that there is a dog in the center, we can't help but see it every time we look. When we hear "Shall I compare thee to a summer's day?" we don't expect a weather report. Given the sentences "The fly buzzed irritatingly on

What's this? Though a familiar image, it may take a while to recognize the dog in the center. This is a good example of the brain's remarkable ability to organize seemingly confusing or obscure information to find a coherent meaning.

the window" and "The man picked up the newspaper," we know what is about to happen.

We take this kind of mental work for granted, but computers can't come even close to doing it. Of course, this may be because our brains have had millions of years to develop, while computers have been around for only forty years. In theory, a computer could someday be created that would perform these mindlike tasks. But a lingering question remains, one which keeps AI researchers awake into the wee hours: Would such a computer necessarily work as our brains do?

The Symbolic Mind

The answer to this question depends on the level at which cognitive researchers examine the processes we use to solve problems and get along in the world. For decades scientists have tried to understand the mind at a symbolic level—that is, starting from the premise that the thinking process involves using rules to manipulate symbols. Such an approach is certainly applicable to examining how we solve the equation 16 x 4 = ? or reason that if all cows are white, and Bessie is a cow, then Bessie must also be white.

Using rules and symbols is a useful way to explore the world around us—it's the heart of physics, for example. If you want to know the force being applied to the bowling ball you are about to roll down a lane, for instance, you can measure the ball's mass and its acceleration. In the rules and symbols of physics, these things are related by the equation $F = ma$. You can manipulate that abstract equation and create something that is still consistent with the real world you are studying. For example, using the rules of algebra, you can transform the formula into $m = F/a$. So if you know the ball's acceleration and the force you apply to it, and want to know the ball's mass, you can use the rules and symbols to calculate that too. Using similar rules and symbols, astronomers determine the mass of things that are not so easily rolled down a bowling lane, such as the moon.

Most people who spend their careers trying to make machines "think" assume that the rules-and-symbols approach to studying the workings of the world also applies to studying the workings of our minds. They believe that if they could discover the right symbols to represent the things we think about and the right rules to manipulate those symbols, then they could put them in a computer, and the machine would think.

At first, it certainly seemed that this approach was correct. Researchers created computers that could often do thinking tasks such as math and logic problems faster and more accurately than our brains could do them. Once these problems were mastered, the researchers moved on to trying to find the rules and symbols that would enable computers to do the thinking tasks we do in everyday life, such as understanding speech and recognizing visual images. Despite twenty-five years of effort, however, these problems have yet to be solved.

This is the kind of text a conventional computer can read:

This is the kind your brain can read:

In fact, digital computers cannot even approach the abilities of a three-year-old child. This failure to account for our everyday cognitive abilities, among other things, has led to a growing suspicion that perhaps the people who brought us "I symbol-process; therefore, I think" might have been putting Descartes before the horse.

It's not unreasonable to conclude that since *some* of our thinking involves using rules to manipulate symbols—and some psychological experiments have suggested this—then at its basic level, *all* types of thinking involve symbol manipulation. But might it not be the other way around? Perhaps most of what we call thinking is something else, something that produces the fuzzy, intuitive, pattern-matching, and prone-to-mistakes stuff that our brains seem to do so easily, and processing symbols is a sideline, more the exception than the rule, a layer of rule-and-symbol icing on the cognitive cake.

A close look at the human mind in action might make you think so. We are awful at doing things that computers do easily and brilliant at doing things that computers have a terrible time with. Where we are intuitive, insightful, and sloppy, computers are methodical, stubbornly literal, and precise. Computer researchers hoping to write programs containing the rules that experts use to make decisions have found that experts often don't have any idea what rules, if any, they use to make decisions. "Our minds are less of a general purpose cognitive device than we might like to think," says Brown University cognitive scientist James A. Anderson. "Complex tasks that are *biologically* relevant—throwing a ball, recognizing a face, understanding speech—are so effortless that we do not realize how hard they are until we try to make a machine do them. On the other hand, the pitiful mess most humans make of logical reasoning or arithmetic would embarrass a $10 pocket calculator."

In other words, using a human brain to do symbol processing may be a little like using the head of a wrench to drive a nail. It might do the job, but that isn't what it's really made for.

But if symbol processing isn't the basis of our brain's ability to think, what is?

Rethinking How We Think

This is the question that a new band of psychologists, physicists, computer scientists, and philosophers is trying to answer. These scientists are trying to create an alternative to the traditional approach to understanding the brain and mind, and testing their new model of the mind on a radically new type of computer. Their revolutionary approach may not seem so unusual to ordinary people: For inspiration on how to build their thinking machines, they are looking to that other thinking machine, the human brain. "Nature is more ingenious than we are," wrote Terry Sejnowski and philosopher Patricia Churchland in a recent paper. "The point is, *evolution has already done it,* so why not learn how that stupendous machine, our brain, actually works?"

The neurobiology involved in this new model of the mind isn't much more sophisticated than a high-school biology course. What is the brain made of? Lots and lots of cellular switches called neurons. What happens when you connect a whole lot of cellular switches together? Somehow, that collection of neurons becomes smart. This intelligence doesn't come from the neurons being fast—they are about a hundred thousand times slower than a typical computer switch. But what a brain lacks in speed it makes up in "wetware," as it's sometimes called. The brain has around one hundred billion neurons, and a single neuron may be connected to as many as a hundred thousand other neurons.

This vast array of interconnected neurons may be what produces the grand collective conspiracy we call our minds. Simple elements often display complicated behavior when they come in large groups. "Suppose you put two molecules of gas in a box," says John Hopfield, a theoretical physicist turned brain and mind researcher at the California Institute of Technology. "They move around the box, and every once in a while they collide. If we put 10 or even 1,000 more molecules in the box, all we get is more collisions. But if we put a billion billion molecules in the box, suddenly there's a new phenomenon—*sound waves.* Nothing in the behavior of two molecules in the box, or ten or 1,000 molecules, would suggest to you that a billion billion molecules would be able to produce sound waves. Sound waves are a collective phenomenon of a complex system."

Might a large group of interconnected neurons display similar collective phenomena? Hopfield's theoretical work suggests they might. In a paper published in the *Proceedings of the National Academy* in 1982, Hopfield showed that a network of neuronlike, interconnected switches could be thought of—and analyzed mathematically—as a physical system. Like the sound waves emerging from the collective interaction of molecules, a group of interconnected neurons have collective properties, and these properties transform a neural network into a thinking machine.

Hopfield's theories have added fuel to the efforts of researchers around the country who are busy trying to find how a neural-net computer might actually take shape. Though they differ in detail, their machines share many of the same attributes. Neural networks are quite unlike conventional digital computers. They are built with components modeled after the neuron and grouped together in interconnected nets. Unlike digital computers, neural nets don't have a "central processor" that operates on a few bits of data at a time. Instead, like our brains, the neurons act on data all at once, bringing the entire system to bear on a problem. Also like our brains, memories in a neural net are spread throughout the network, not housed in a separate memory bank.

Neural nets also process information differently from digital computers. Each neuron takes in signals from the other neuronlike components, adds them up, and decides on the basis of the answer whether to send out a signal of its own. In a way, the neural units in the machines are analogous to people in a jury talking among themselves, trying to influence each other to decide one way or another; when an input comes into a neural network, there is a mad jumble of changing votes and opinions at first, but eventually all the neurons settle on a decision and the machine produces an answer.

Another important difference between neural networks and digital computers is how they are programmed. Computers are typically given a list of instructions to follow. A neural network, on the other hand, is "taught" through a series of examples. A layer of input neurons is shown a pattern—a grid of light and dark squares, for instance—and the connections between the input and output neurons are modified to produce the correct answer—a signal indicating, for example, whether the pattern of squares is symmetrical. After being exposed to many patterns and being "told" by the programmer whether they are symmetrical, the machine eventually learns to make the symmetry judgments for itself, even on patterns it hasn't seen before. Biophysicist Terry Sejnowski has designed a neural net that does just that.

Not only are neural networks programmed differently from computers; the answers they produce are a little different as well. A neural network's freewheeling style of information processing doesn't always produce an exact answer, but neural nets can solve some kinds of problems faster than conventional computers and produce answers that are almost always *very good*. With some tasks, such as pattern matching, it is often more important to find a good answer quickly than to get the answer exactly right. For example, in trying to decide whether the creature running towards you is a potential danger or a potential dinner, speed is as important as accuracy; it's better to get out of there fast than to precisely identify the animal as a Bengal or a Burmese tiger.

Like our brains, neural nets can draw inferences. "If you hear the words *bat, ball,* and *diamond,* you think of one thing," says Brown University's Anderson, who uses neural nets to study how we make categories, "and if you hear the words *bat, vampire,* and *blood,* you think of another." Given *bat* or *diamond* alone, Anderson's neural net will respond with characteristic qualities of animals or geometric shapes. But if *bat* and *diamond* are put together, the machine comes up with *baseball.* While it is also possible to program a conventional computer to draw these kinds of inferences, making such distinctions is a natural property of neural networks; because memories are stored throughout the entire network of units, common traits overlap in the connections.

A neural network's ability to form categories and make associations provides cognitive scientists with a new tool for studying how we think. Researchers have created a neural net that makes generalizations about the relationships between members of a family; another network is learning to change the tense of verbs and displaying the same type of successes and mistakes that children do when they learn a language. Neural nets have also piqued the more practical interest of governmental research labs and commercial companies, many of which are beginning to explore how such networks might be made into silicon chips.

It's unlikely, however, that neural nets will completely replace the good old number-crunching computers that we've grown so used to. As proponents of traditional AI like to point out, relying on machines that think too much as we do may not be such a great idea. You certainly wouldn't want a neural network computer, with its inexact calculations, to balance your checking account or do a company payroll.

The impact of neural nets, however, may go far beyond simplistic promises of creating new kinds of computers. Rather, neural networks represent a fundamental shift in science, a basic change in the way cognitive researchers are approaching the study of our minds. Instead of looking for the rules by which our behavior is governed, neural networkers are exploring how groups of interacting neurons in the brain might give rise to those behaviors—and perhaps even how groups of interacting brains can give rise to the social and anthropological features of human societies.

Neural network models may become a common ground for scientists who study the far corners of the brain and mind. "The power of neural networks comes from their simplicity," says Gary Lynch, a neuroscientist at the University of California at Irvine. "Researchers in many different kinds of disciplines can understand them, and that makes it easier for us to understand each other." Terry Sejnowski agrees: "Neural nets are not just a way of getting new answers to old problems. They will also produce new questions. The models represent a new language in which different kinds of researchers can talk about the mind and brain."

Not all researchers are so enthusiastic about the neural network approach to understanding the brain and mind. Scientific theories are mountains of sand built grain by grain, and people in the mountain-building business are justifiably wary of anybody who comes their way driving a bulldozer. The ranks of cognitive scientists are populated with researchers who overthrew their predecessors' established theories, and they are unlikely to quickly abandon their hard-earned theories for something new. Yet these researchers are also deeply interested in how the brain and mind work; therefore, many are content to watch at a distance, half hoping that the neural networkers are on to something.

The ranks of neural net revolutionaries are filled with people from every discipline and training: neuroscientists, psychologists, computer scientists, linguists, physicists, and even philosophers. Together they form a group that at first might resemble the central casting call for a Hollywood epic film—perhaps a bizarre version of *The Dirty Dozen*—but this time the assault is on the mind and brain. On a poster for the movie, the descriptions of the main characters might read something like this:

- Jim Anderson, the "old guard," a pioneer who began working with the new model twenty years ago, when it was unfashionable, and helped nurse it to its present stage.
- John Hopfield, the eminent theoretical physicist turned brain researcher, who helped spread the word about neural nets by showing how they might be made into machines.
- Gary Lynch, a cigar-chomping neuroscientist who gets his hands dirty in the lab studying our sense of smell and uses neural nets to ferret out the brain pathways of our senses.
- Patricia Churchland, a Canadian philosopher who went to medical school to study how neurons work and now is bringing the brain back to philosophy—threatening the fundamental tenets of her profession as she does so.
- George Lakoff, a linguist who found his way to the new model almost by accident, whose work is challenging the theories of Noam Chomsky, Lakoff's former teacher and the most influential linguist of the century.
- David Rumelhart and Jay McClelland, the "new" old guard. These two cognitive psychologists are part of the backbone of the recent resurgence of the new model of the mind. They have developed new models for neural nets and new mathematics for training them; they have shown how neural nets can be useful in studying how we think. Their three-volume book on neural nets is the handbook for neural networkers around the world.
- Geoffrey Hinton, an eccentric Englishman and expert in artificial intelligence. Brash and witty, Hinton is one of the field's most eloquent defenders; his theoretical research helped provide a breakthrough in training neural nets.
- Terry Sejnowski, the next wave. A physicist turned neuroscientist, he represents a new generation of neural net revolutionaries. As a graduate student at Princeton University, he wrote one of the first Ph.D. dissertations on the behavior of neural nets. As a professor of biophysics at Johns Hopkins, he oversees projects in neuroscience, artificial intelligence, and psychology. He is betting his career that the neural net theories will prove to be correct.

These researchers have been joined by many more scientists, ranging from Nobel Prize winners like Francis Crick and Leon Cooper to graduate students in cognitive and computer science. They are taking on the next great project in science—unraveling the most complex thing in the universe, our brain. They are approaching this goal in a revolutionary way, exploring how machines modeled after our brains might generate mindlike behavior. The researchers are not so much trying to take our minds apart as they are trying to put one together.

They call themselves connectionists.

CHAPTER TWO

THE PUZZLE MACHINE

The Mechanics of Thought

Imagination is more important than knowledge.
—Albert Einstein

The real danger is not that computers will begin to think
like men, but that men will begin to think like computers.
—Sydney J. Harris

"Well, the first thing you see is the outside triangle, so you figure
that it's possibly that one because it's real different on the outside.
Then you realize that many of them are different on the inside."
Jay McClelland, sitting in his office at Pittsburgh's Carnegie-Mellon
University, is trying to solve a seemingly simple problem. The
forty-year-old professor of psychology is looking at the figures
shown below.

The question is, Which of these figures is "most different"?
"Let's see," he says. "Well, the triangle is different. But why
should differences on the outside be any more important than
differences in the inside? So let's look for some more differences.
There are several: whether the outside is a circle or not, whether
the inside is a circle, whether there's a dot in the middle of the
inside, and whether the inside is standard size. On that basis, I'd
say that the first one is sort of the prototype, because it has the

typical value on all the dimensions. And the second one differs on one feature, the third one differs on one feature, the fourth one differs on one feature, and the fifth differs on one feature. So on that basis I would say that I can't make a decision."

This problem, designed by psychologist Peter Hayes, is tricky. The first figure, which McClelland called the prototype, shares the most features with all the other figures. It is the least different of the group. But since all the other figures differ by one attribute, this "least different" figure is in fact the "most different" because it is the one that is most alike. On being told this solution, McClelland reacts the way most people do—with a Bronx cheer.

Put an index finger on each temple. Between your fingertips lies a three-pound blob of biology that is an incredible machine. It does mundane regulatory tasks like keeping you upright and breathing. It does sophisticated motor tasks like guiding your fingers to get an egg from the refrigerator and crack it into a hot frying pan. It tells you what the world out there is like: picking up molecules in the air that indicate the toast is burning and sensing the airborne shock waves that tell you the cat is meowing to be fed. It synthesizes information, too: It knows that the objects on the kitchen shelves are glasses, not coffee mugs, and that the things in the sink are bowls, not plates. Your brain would probably know all this even if it were the first time it had seen these things.

Your brain doesn't just process information; it also stores it in a number of ways. You can look at a number like 244–7628 and remember it long enough to get from a phone book to a telephone, but if the line is busy, you may have to look up the number again to redial. If 244–7628 is your best friend's phone number, you probably wouldn't have to look it up at all. And if you think hard enough, you can probably still remember a phone number you had when you were a child.

The information the brain is capable of storing goes beyond simple numbers: a whiff of perfume, the smoke from a pipe, the smell of hot tar in an August afternoon, the waft of waffles—all can cause your brain to spill reels of memories of a loved one, a cherished ritual, or even just a simple activity from long ago. Often you remember more than visual scenes; the brain can evoke the eerie sensation of the way you *felt,* too. Think of one of your most embarrassing moments; even though it was long ago, you may still feel an uneasiness in your stomach.

The mind also creates new ways of manipulating the environment around it—albeit not always for the better. The problems it solves go beyond simply building better and better mousetraps. In fact, it tries just about anything. The same brain that invented the wheel and the digital watch also tries to solve less practical puzzles such as *How did the universe begin?* or *What is morality?* In fact, this same brain actually *enjoys* doing problems just for fun—puzzles such as *Which of these shapes is most different?*

How do our brains do all these things? For the last three decades, most cognitive scientists have researched this question under a basic assumption: If you observe the human brain in action long enough, you'll notice that it displays some kinds of regular, mechanistic behavior that can often be characterized as a set of rules. But unlike auto mechanics, brain scientists can't examine the machinery of the mind by looking under the hood. So instead they "test-drive" the brain by giving it puzzles.

Putting the Brain to Work

Some of the puzzles that brain researchers ask their subjects to solve are not really puzzles in the ordinary sense. Many are so simple that people do them almost instantaneously and without consciously thinking. In one type of study, for example, subjects sitting in front of a screen are presented with pairs of letters such as *aa* or *ac*. They are asked to push one of two buttons, indicating whether the two letters are the same or different. It sounds pretty easy, but when, for example, the pair *Aa* is presented, it takes longer to decide whether the example contains the same letters than when *aa* is shown. Participants are not consciously aware of the delay, and it is not much, less than a tenth of a second, but it happens repeatedly. This "reaction time" (how long from the presentation of a problem to the solving of it) can be a clue to what's going on in the brain.

In another example, people are shown figures like those on the following page and asked whether they are identical, except that one has been rotated in space. There is evidence that the more a figure has been rotated, the longer it takes to recognize it as identical to the other figure.

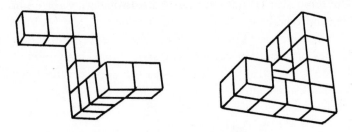

Are these two figures different views of the same object?

Of course, this approach to understanding the brain is a little like trying to determine how an automobile works by simply driving it around. Pushing the accelerator down, slamming on the brakes, zipping it around curves—by watching how a car performs different tasks, you might gain some insight into how it works. If you experiment long enough, for example, you might hit upon the relation between an auto's performance and the presence of gasoline in the tank.

In the same way, test-driving the brain clearly demonstrates that when we solve problems, *something* is going on inside our skulls, even though we may not be conscious of it. In the first experiment, the brain is simply matching figures when it sees *aa*, but when it sees *Aa*, it is not only comparing the figures' shapes but also comparing their *meaning*, and that takes longer. Nobody knows quite what to make of the studies of the rotating figures. Conducted by Stanford University psychologist Roger Shepard and Ulric Neisser and Lynn Cooper, both at Cornell University, the studies are often interpreted as demonstrating that the brain can think in images instead of words or symbols.

While there are reams of reaction-time data, psychologists can only speculate what is going on inside the "black box" of our brains while it solves these kinds of problems.

Therefore, some psychologists use other kinds of puzzles to try to find those rules that govern the mind's machinery. Subjects are given a puzzle and asked to say out loud what is going on in their minds while they are trying to solve it—just what McClelland was doing with the "most different" problem. In one experiment, for example, a student was asked to say what was going through his mind as he solved this cryptoarithmetic problem:

Knowing that D = 5, what numbers do the other letters in this problem represent?[1]

$$\begin{array}{r} \text{DONALD} \\ + \text{GERALD} \\ \hline \text{ROBERT} \end{array}$$

The student's comments while he solved the problem, detailed in a book by computer researchers Allen Newell and Herbert Simon, *Human Problem Solving*, ran some fifteen pages. To solve the problem, the student showed a variety of rulelike behaviors: setting goals, making assumptions and testing them, and rejecting hypotheses that didn't work.

Using similar types of rules, Newell, Simon, and Cliff Shaw created a computer program called the General Problem Solver, or GPS. It could look back at its performance and look forward to its goals, generating possible strategies and testing them. It could prove theorems in formal logic, do trigonometry, and even solve puzzles similar to this one:

> Three cannibals and three missionaries are on one side of a river and need to get to the other side. There is a boat at the bank of the river, but it is small and can hold only two people at a time. How do you get all the cannibals and missionaries across the river without ever allowing the cannibals waiting on either side of the bank to outnumber the missionaries?[2]

[1] $\begin{array}{r} \text{DONALD} = 526485 \\ + \text{GERALD} \ = 197485 \\ \hline \text{ROBERT} \ = 723970 \end{array}$

[2] The cannibals and missionaries puzzle: A missionary and a cannibal cross the river together first, then a missionary returns alone. Next, two cannibals cross together, and a cannibal returns alone. Then two missionaries cross together, and a cannibal *and* a missionary return together. Then two missionaries cross, a cannibal returns, then two cannibals cross, one returns, and the last two cross.

The GPS was very powerful because a problem like the canni-bals and missionaries could be solved using rules, in this case the rules of logic.

Rulelike behavior isn't found only in problem solving. It can be found in simpler cognitive processes as well. Suppose you were in a room that was totally dark, except for a single spotlight shining on a sheet of paper hanging from the ceiling. No matter what color the paper is—suppose it's dark gray—the sheet will appear white to you. But if you put that same piece of paper in a well-lighted room, it will show its true color. The light that reflects from the paper to your eye is the same regardless of whether the paper is in the dark or in a lighted room. But the paper *looks* brighter if it is surrounded by darkness. In other words, your perception of the brightness of the paper depends on the intensity of light surrounding it.

The brain might accomplish this effect by using a rule such as *When you see a patch of light, judge the brightness of that patch of light by comparing it to the brightness of the areas surrounding it.* Such a rule might be ideal for a conventional computer. You can imagine a computer, hooked up with an array of light sensors, whose central processor, armed with the "look up, down, left, and right" rule, simply goes down the line and checks each sensor. The computer's "eyes" respond like our eyes.

This rule-based model of the mind was extremely appealing to artificial intelligence researchers and psychologists. Computers are *designed* to be run by rules, making them perfect for testing rule-based theories of how the mind works. To psychologists, explor-ing the mind of a "higher" level of rules seemed a plausible way to avoid having to consider the complex interaction of the brain's billions of neurons.

Bringing the Brain Back In

Plausible as this approach might appear, however, McClelland and the other connectionists are trying to replace this rule-driven descrip-tion of our thinking processes with a new model of the mind that incorporates the structure of the brain itself. In the neural network model of the mind, thinking is not following a set of rules but a product of the complex interactions of huge numbers of neurons.

For example, another way of understanding brightness perception, says McClelland, is to use a neural network of light-detecting neurons connected to each other in such a way that if any one neuron is stimulated by light, it inhibits the sensitivity of other neurons around it. If the network sees the piece of paper in a bright room, all the neurons in the network are excited and therefore inhibit each other; the paper appears less bright. But if a small group of neurons is stimulated by light while the surrounding neurons are in darkness—which is what happens when an illuminated piece of paper is hanging in a dark room—the input from those excited neurons will be greater because they are not inhibited by the neurons in darkness.

In fact, that's exactly what happens in your eyes. Your retina is a network of mutually inhibitory light detectors; your eye's neural net performs according to a rule, but the rule isn't actually "in there" anywhere. "What I, as a connectionist, would urge everybody to do is to separate the rule that you write down to *describe* something's behavior from the question of how the behavior is actually *implemented,*" says McClelland. "Being born with a mechanism that acts in accordance with some rule doesn't mean you are born with a rule that's in your head."

Mistaking a description of a phenomenon for an explanation of how it works isn't confined to cognitive science. The model of the Solar System proposed by the Alexandrian astronomer Ptolemy, for example, put the earth at the center with the Sun, Moon, planets, and stars whirling around it. Developed during the second century, the model was quite good at predicting all the planets' behavior, though as observational techniques improved, the model had to be revised with more and more outlandish embellishments. For example, ancient astronomers puzzled over why Mars sometimes seemed to stop in its tracks, move backward, then go forward again. Astronomers now know that Mars's *retrograde* motion is due to the fact that the Earth travels faster in its orbit than its neighboring planet. As the Earth approaches and passes Mars, the red planet appears to slow down, stop, then go backward—just what someone in a car passing another car on the freeway would witness. Once the Earth travels around its orbit a bit, Mars appears to move forward again.

To correct for these discrepancies in Ptolemy's description of planetary motion, a tiny circular orbit was added to Mars's normal orbit around the Earth. Every once in a while Mars would spin around this smaller orbit, going backward, then forward again. Though at first Ptolemy's model was intended simply to aid in predicting the motion of the planets, by the time the model was finally overturned by Copernicus's Sun-centered model in the sixteenth century, Ptolemy's *description* of the planets' motions had become widely accepted as the model that *explained* the planets' motions.

In the same way, Einstein's theory of relativity showed that in fact, Newton's laws of gravity are only an approximate description of bodies in motion. At velocities approaching that of light, time and space become warped, and Newton's laws aren't accurate. Still, Newtonian mechanics remains extremely useful to physicists trying to understand simple phenomena in the everyday world, because physicists know at which point that the Newtonian description of the world is inaccurate.

In this regard physicists have a significant advantage over cognitive scientists; mind researchers don't really know where their rule-based understanding of the mind falls apart. "It's got to be useful to talk about explicit rules in behavior," says McClelland. "When I write the word *receive,* for example, I recall this jingle: '*i* before *e,* except after *c,*' and then I know what to do. But at what point do the approximations break down in cognitive science? For a long time we thought that the *rule* level of cognition was sufficient. But it's not."

One of the primary reasons for connectionists' rejecting making rules to approximate the activity of the mind is the same reason rules became popular in the first place—the regularity of our behavior. At some level, it does appear that we behave in a rulelike manner. But if you look closer, you see quite the opposite; it seems that we regularly use anything *but* rules to get by in the cognitive world.

Instead, we appear to use a process closer to intuition. For example, most of us have been doing "most-different" problems since the first time a child psychologist waved an intelligence test under our noses. Since the problem shown at the beginning of this chapter used a similar language, format, and shapes to other puzzles, McClelland naturally assumed it was no different. You could almost say that he actually behaved brilliantly because he took cues from the way the problem was presented, immediately generalized them to other problems he had seen, and realized what kind of information he needed in order to find a solution. This generalization process is a large part of solving any problem.

By giving a Bronx cheer to the solution to the "most-different" problem, McClelland follows a long tradition of problem solvers who realize that sometimes the biggest obstacle to doing problems like these is their own brilliant, intuitive brain. This particular problem takes unfair advantage of our brain's remarkable ability to make generalizations. For most problems of this type, determining differences among shapes, numbers, or letters is the key to a solution. But the problem at the chapter's beginning takes the task one step further: The hardest part is not looking for differences, but realizing that you have to reject your initial assumption of what "most different" means. Once you do that, the rest is easy.

There are other problems more brazen in their challenge to our brain's powers of assumption: *How many types of animals did Moses take on the ark? If a plane crashes on the U.S.-Canadian border, in which country are the survivors buried? Is it legal for a man to marry his widow's sister?*

These problems are mostly designed not for the entertainment of the solver, but for that of the asker who can smirk while you struggle to "solve" the problem instead of realizing that Moses didn't go on the ark, that survivors don't get buried, and that a man who has a widow is dead.

Such problems are effective because when someone says, "Hey, I've got one for you," our brains click on, ready to soak up as much information as possible. As the brain takes in that information, it also sifts through it, searching for the parts that are relevant to solving the problem and throwing away those parts it considers extraneous. Because it does all this very quickly, the idea of Moses on the ark sometimes slips through. Moses has enough of the characteristics of Noah to satisfy the brain that there's a biblical figure on an ark, and so it moves on to more important information. The brain's no patsy, though. Try asking, "How many animals did Nixon take on the ark with him?" and see how long it takes somebody to figure out what's going on.

Much of our mind's power comes from the fact that it's a good guesser, and problems like these turn that power upside down. Trick problems would be easy for a computer because they are basically asking just for factual information with a red herring thrown in. It would take just as long for a computer to look up a description of Moses or Nixon in its memory and respond, "This does not compute." Of course, if someone seriously asked you what book in the Bible told the story of Moses and the Ark, you might respond, "It's in Genesis—and by the way, it's Noah, not Moses." A computer would still respond, "This does not compute." A computer isn't tricked by trick questions because, in a way, it isn't smart enough.

These kinds of problems are relatively rare because, like judo masters, they use the great strength of their opponent—the brain—to their own advantage, and that requires some skill. Also, such problems may be rare because there is a far easier way to throw a banana peel in the brain's lumbering path. (See the Figure below.)

If one, and only one, of the inscriptions on the boxes is true, which box should you open to find the treasure?

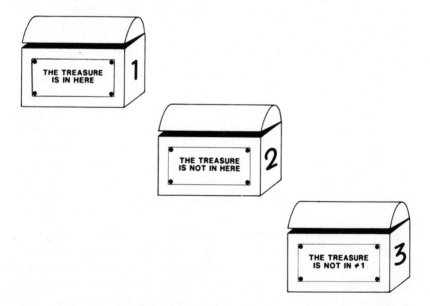

Answer: Some of the most logical puzzles are in logical terms, fairly simple, and would be easy for a computer to solve. But they are a challenge for the less logically-minded human brain. The only way one statement on a box can be true while the other two statements are false is if the treasure is in box number two.

In fact, typical puzzle books are filled with problems of this type, which in other forms appear thus:

> Four men and four women are shipwrecked on a desert island. Eventually each one falls in love with another, and is also the object of one other person's love. John falls in love with a girl who unfortunately is in love with Jim. Arthur loves a girl who loves the man who loves Ellen. Mary is loved by the man who is loved by the girl who is loved by Bruce. Gloria hates Bruce and is hated by the man whom Hazel loves. Who loves Arthur?[3]

The reason these types of puzzles are so numerous—and so difficult for us—is that they go straight for the Achilles heel, as it were, of the brain: using logic.

Our Irrational Mind

In cognitive science, thinking logically does not mean being sensible or reasonable. It means following the rules of formal logic. Throughout this century philosophers have worked hard to develop the rules that govern logical thought in what is called deductive reasoning. For example, complete this syllogism:

> Everyone at the picnic is an artist.
> Mary is at the picnic.
> Therefore, . . .

You probably solved this one without consciously trying, and You could say that your thinking followed the rules of logic. The syllogism could be thought of as a formula:

> All p's (picnic-goers) are a's (artists).
> m (Mary) is a p.

Using the simple rules of formal logic you get, "Therefore, m is also an a."

No matter whether the subject is Mary or Mike, picnics or

[3]Gloria loves Arthur.

poolrooms, artists or acrobats, once you have the formula, it works for anything or anybody.

Because such reasoning problems can be reduced to formulaic rules, logic can be very powerful. For instance, it can help do syllogisms whose conclusions are not quite as evident as the previous one's. For example:

> Some of the beekeepers are artists.
> None of the chemists are beekeepers.
> Therefore, . . .

Stumped? You're not alone. Philip N. Johnson-Laird, a British psychologist, found that more than half the people given this syllogism said that there was no logical conclusion they could draw from it. There is, but it might take somebody trained in logic to find it. The conclusion is, *some of the artists are not chemists.*

If it's difficult to make sense of the previous syllogism, try another with the same format but different players:

> Some mammals can fly.
> No birds are mammals.
> Therefore, there are some flying animals that are not birds.

For most people this syllogism is easier to grasp. But for a computer, the task of evaluating the beekeepers-and-artists and the birds-and-mammals syllogisms is equally easy, because the elements and relationships of both syllogisms can be reduced to a formula. A computer doesn't need to take into consideration what a syllogism is about, so it can simply apply the rules in its program to a symbolic representation of the problem.

The ability of computers to tackle these difficult logic problems with ease is partly why many people became overly optimistic about—or fearful of—the power of digital computers when the machines were being developed in the 1950s and 1960s. For decades, researchers have assumed that at some fundamental level, both brains and computers worked the same way. Since math and logic problems were considered to be things that smart people could do better than stupid people, the computer scientists thought the same thing about their machines. In a sense, they thought that they were working backwards: If computers could easily do the kinds of problems our brains have such difficulty with, then doing the things our brains do easily—understanding speech, for example— would be easy for the machines.

It hasn't turned out that way. The first programs in artificial intelligence could do college-level calculus, observes computer pioneer Marvin Minsky, of the Massachusetts Institute of Technology (MIT). But when programmers tackled high-school algebra they found to their surprise that it was much *harder* to do on a machine. Grade-school mathematics—the concept of numbers—proved an even greater challenge. And exploring the child's world of blocks was almost insurmountable, except within very narrow limitations. It finally dawned on AI researchers, says Minsky, that most of what we call intelligence is developed within the first year of a child's life.

The research of Johnson-Laird gives a clue to why scientists are having such a difficult time making computers think like a brain, for it suggests that the brain may be more than a somewhat imperfect logic machine. For example, most people find that some syllogisms are easier to understand than others, even if they are logically identical, and that's something of a puzzle in itself. If our cognitive powers are based on applying rules like those of formal logic, it would seem that structurally identical puzzles should be equally easy to understand. But they're not. Here's a syllogism from psychologist John Horn of the University of Denver:

> No Gox box when in purple socks.
> Jocks is a Gox wearing purple socks.
> Therefore, Jocks does not now box.

It might be difficult for some people to tell whether the conclusion is valid. But try this one, which has the same structure but a more familiar subject:

> No authors write while playing the saxophone.
> Mike, an author, is playing the saxophone.
> Therefore, Mike is not now writing.

In this case, our familiarity with authors, writing, and saxophones seems to help us determine that the conclusion is valid. In fact, for some syllogisms, we appear to reject logic outright and rely *only* on our familiarity with the subject matter to determine whether it is valid. For example, consider this syllogism:

> Democratic nations believe in free speech.
> The people of the United States believe in free speech.
> Therefore, the United States is a democratic nation.

No matter how reasonable this conclusion seems, it is not logically valid. Of course, the conclusion may be valid for a lot of other reasons, but as for the syllogism itself, it is structurally no different from this one:

> Items of exceptional quality are expensive.
> This item is expensive.
> Therefore, it must be of exceptional quality.

Now try this:

> A low population growth is beneficial to a nation.
> Famine lowers the population growth of a nation.
> Therefore, famine is beneficial for a nation.

Despite its grim, "unreasonable" conclusion, this syllogism is logically valid.

These syllogisms suggest that when we reason, we don't rely solely on logical rules but depend much more on the content of the problem, applying some other kind of reasoning. It's not clear what this other kind of reasoning is, but connectionists feel confident about one aspect of it: "The laws of thought," says McClelland, "are not the laws of logic."

There is further evidence of just how little we rely on logic when we reason. In a study by psychologists Daniel Kahneman, Paul Slovic, and Amos Tversky, subjects were given a short description of a person and then asked to guess which professions and hobbies the person was most likely to have. Here's one of their examples:

> Russ is 34 years old. He is intelligent, but unimagina-tive, compulsive, and generally lifeless. In school, he was strong in mathematics but weak in social studies and humanities.

Please rank in order the following statements by their probability, using 1 for the *most* probable and 8 for the *least* probable.

Russ is a physician who plays poker for a hobby.
Russ is an architect.
Russ is an accountant.
Russ plays jazz for a hobby.
Russ surfs for a hobby.
Russ is a writer.
Russ is an accountant who plays jazz for a hobby.
Russ climbs mountains for a hobby.

Most people, quite reasonably, ranked "Russ is an accountant" as most probable. They ranked "Russ plays jazz for a hobby" as very unlikely. However, most people also said that the probability that "Russ is an accountant who plays jazz for a hobby" is higher than the probability that "Russ plays jazz for a hobby."

Ranking "Russ is an accountant who plays jazz for a hobby" as more probable than "Russ plays jazz for a hobby" violates the laws of probability. It is impossible for a statement combining two unrelated elements—"tomorrow it will rain, and I will get a letter from my mother"—to be more probable than either element alone: "It will rain tomorrow" or "I will get a letter from my mother tomorrow."

Therefore, the chances that "Russ is an accountant *and* plays jazz for a hobby" can't be greater than the chance that "Russ plays jazz for a hobby." Yet most people in the study said that they were.

Perhaps you might expect that result from people who had no training in probability and statistics. But the researchers used the same test on a group of graduate students in psychology and medicine who had taken courses in statistics. They also tested a group of students in a "decision science" program at the Stanford Business School who had taken advanced courses in probability and statistics. In all groups, more than 80 percent of the students ranked the probabilities incorrectly.

These studies may cast our mental abilities in an unflattering light, but in a way, they also demonstrate our brains at their best: Russ's personality profile so strongly matches the stereotype of an accountant, and so strongly mismatches that of a jazz player, that we simply stick with our impressions based on the profile and ignore the rules of probability and statistics. Choosing "Russ is an accountant who plays jazz for a hobby" as more likely than "Russ plays jazz for a hobby" is incorrect according to the rules of probability, but it demonstrates our brain's ability to consider the wealth of other information that Russ's personality profile is subtly giving us.

In some instances we ignore the rules of logic outright. Suppose, for example, you buy and pay for two theater tickets that cost $20 each. When you arrive at the theater on the night of the production, you discover that you can't find the tickets. You suddenly remember to your despair that you left them in a pocket of a coat you sent to the dry cleaner several days ago. The tickets are gone forever, and there are no refunds, but there is a pair of tickets in the same section of the theater still available for the same price. Do you spend another $40 for a pair of tickets? In one study, most people given this scenario said no.

Now suppose another scenario. Instead of having received the tickets, you are to pick them up and pay for them the night of the show. You arrive at the theater and discover that while you are sure you had four $20 bills in your coat pocket that afternoon, you now have only two. You discover to your dismay that you have a hole in that pocket and that sometime during the evening some money must have fallen out. In any event, you're sure the money is gone, and you are out $40. Now, would you use the remaining money to pick up your tickets for the theater? Most people in the study said that they would.

In terms of simple arithmetic, you have lost $40 in either scenario. And in both cases, you are given the opportunity to buy a pair of tickets. But in the first case, most people said they would not buy tickets; in the other, they said they would. The difference, apparently, is that people tend to keep a mental accounting of their expenditures. Since they had already paid $40 for tickets in the first scenario, buying new tickets would mean, according to their accounting, that they had spent $80 on tickets—to them, too much to pay to see a show whose tickets cost $20 each. In the other scenario, the lost $40 had not yet been committed; it came, according to the mental accounting, from a "general fund." Consequently, they considered themselves to be spending only $40 for the tickets.

A similar type of accounting makes most of us run across town to pay $15 for a radio selling for $30 in a nearby store when we wouldn't make the same trip to buy a television set for $615 when there is one nearby for $630. In either case you are saving $15, but in the second the savings seem much smaller because it's considered in light of the total amount you are spending on the television.

The Insight Machine

The point of these examples is not that we aren't very good at reasoning, though in some cases our inability to handle simple logic can cause us to make a mess of things; rather, these studies suggest that we are doing something else besides applying logic when we reason. That "something else," say connectionists like McClelland, may be responsible for what makes us so smart, in the same way that neural networks (which can't do logic, either) can perform other sophisticated cognitive tasks. In fact, our inability to do logic may be responsible for our mind's remarkable powers. "What makes our brains bad at doing logic tasks is actually the *essence* of our higher-level cognitive functions," says McClelland. "With a connectionist type of machine, it's easy to recognize your mother and hard to do logic. With a conventional computer, it's hard to recognize your mother and easy to do logic. So what does it mean to be *really* intelligent?"

According to McClelland, being intelligent goes beyond simply doing logic or recognizing patterns. Instead, the essence of high-level thinking is making generalizations from past experiences and applying them to new situations. "What's smart is having *insights*," says McClelland. "Being able to figure out what to do in novel circumstances. There's no such thing as insight in a computer program like Simon and Newell's General Problem Solver. It's just mechanically going all through this processing, comparing the current state to the goal state, setting sub goals, all that kind of stuff, in an extremely mechanical way. And yet even when people are doing the same kinds of problems as the computer is doing, they occasionally have insights—they jump way ahead of all this mechanistic GPS-type stuff. In much more complicated kinds of problems, where the ways of approaching the problem aren't even mapped out, there's just no way some GPS-like machine is ever going to work. It just seems to me that whatever makes people able to have insights, is something that's very, very different from rule-based processing."

What makes people create innovative solutions to problems? Those flashes of insight that seem to start as a dim hunch and quickly swell into what psychologists call the "aha!" reaction? Nobody knows. Historians of science point out that often theoretical breakthroughs occur to scientists when they are not actively thinking about a problem. In one famous example, the nineteenth-century German chemist Friedrich August Kekule von Stradonitz was relaxing before a fire and then, as he later wrote, "sank in a reverie. Atoms danced before my eyes. Long chains were firmly joined, all winding and turning with snakelike motion. Suddenly, one of the serpents caught its own tail and the ring thus formed whirled before my eyes. I woke immediately and worked on the consequences the rest of the night." Those consequences resulted in his discovery of the ringlike structure of benzene, an insight that gave birth to modern organic chemistry.

It is very difficult to understand the mechanisms of insight because *insight* is almost *defined* as the ability to get around the conventional mechanisms of solving a problem. Many "insight" problems have other solutions that can be hammered out through trial and error or a series of laborious steps. In a sense, insight is *illogical;* instead of taking a direct route to an answer, you go in another direction and come up with an even more direct route.

Here's a classic insight problem:

> One morning at sunrise a hiker starts his way up a narrow path to the top of a mountain. He walks at various speeds, stops for lunch along the path, and finally reaches the top at dusk. He spends the night in a cabin at the top of the mountain, and the next morning at sunrise he begins his way down the mountain path. Again, he walks at various speeds and stops for lunch. He gets to the bottom of the mountain in a little less time than it took him to get to the top. Will there be a spot along the mountain path that the hiker occupied at exactly the same time on both days?

To solve this problem you might try struggling with the mathematics of the average speed the hiker traveled and his position on the mountain during the hike. But the more insightful solution is to imagine that the hiker has a twin brother who starts on the path going up the mountain at the same time the first hiker leaves to go down. At some point along the way, the two walkers would have to cross each other's path, meaning that the hiker going down must cross the path of his own upward journey at the same time both days.

This kind of puzzle is an example of a problem that has both a slow, straightforward solution and a quick one that requires insight. The difference between solving a problem using insight and solving it with methodical, rule-based thinking is nicely illustrated by watching a human play chess against a computer. Computer chess programs were first created years ago as an attempt to study the human mind. It was thought that creating a chess-playing computer would be a good way of studying how humans solve problems.

In the early days of artificial intelligence computers were slow, so chess-playing programs relied on knowledge about chess strategy to select a few possible moves. Then the program checked out the possible moves by "looking ahead"—playing each move, and the possible counter-moves an opponent could make, and the possible responses to that response, etc.—and chose the best option.

But these programs were prone to making mistakes. As computer hardware got faster and faster, looking ahead at many more of the millions of potential moves and counter-moves in a typical game of chess became easier and easier. Nowadays, some computer chess programs are ranked among the best chess players in the world, and have even overturned a few tenants of conventional chess wisdom. Ken Thompson of Bell Labs, for instance, used a computer to play out all the possible endgames with five or fewer pieces on the board and found that contrary to what most experts assumed, two Bishops can beat a lone Knight more than 90 percent of the time.

But these modern chess programs, while quite accomplished, for the most part have abandoned the strategy of trying to mimic the human mind. Instead, the programs rely on what programmers call *brute force*. They use some overall chess knowledge, but their strength lies in being able to examine mechanistically hundreds of thousands of chess moves per second.

Though these computer programs play chess very well, their information processing isn't at all like what human players do when they play chess. Humans tend to look very carefully at a few potential moves. How they know which moves to consider may have something to do with the human brain's remarkable ability to learn from experience. In one study, two groups of people—one consisting of chess players, the other of people who had never played chess—were briefly shown a random arrangement of chess pieces on a board. When later asked to reconstruct the arrangement of pieces, the groups were equally adept. Then the groups were shown arrangements taken from records of actual games. This time, the chess players were much better than the non-chess players at reconstructing the positions. The fact that the chess players had seen many games of chess previously apparently made the difference and may explain why human chess players can beat the best chess programs. The players can apparently look at the arrangement of pieces on the board, recognize the pattern as similar to one of the thousands seen in previous games, and make a move based on that experience.

There is an important distinction, however, between matching the position of pieces to a previous experience and recalling a pattern. Chess experts have prodigious memories, but they don't just remember a new pattern of pieces as something they'd seen in a previous game; after all, the pieces may not be in exactly the same places. Instead, they are doing something subtler, something that lies at the very heart of our ability to think. "When you look at an arrangement of chess pieces," says McClelland, "you're synthesizing a response to a new situation that is consistent with the responses you synthesized to related situations."

Producing a new response from previous experiences enables neural networks to do some kinds of cognitive feats without explicitly using rules. McClelland and David Rumelhart, a connectionist at Stanford University, designed a neural network that learned to change the present-tense form of a verb to its past-tense form. Rather than being programmed with rules for constructing the new form of the verb, the network was trained by being shown a group of verbs and their past tenses.

The network memorized irregular verbs—changing *see* to *saw*, for example. But the network was also able to learn various overall patterns for regular verbs—changing *guard,* for example, to *guarded.* Though it wasn't designed to do so, this neural network made the same kinds of mistakes that children make when they are learning verbs. For example, at first the network would learn that an irregular verb such as *go* changes to *went.* But as the network continued learning and assimilated the patterns of regular verbs, it went through a period of changing *go* to *goed.* The phenomenon, called overgeneralizing, has also been observed in children.

Once the network was fully trained, it knew how to transform both regular and irregular verbs, even those it hadn't seen before. "The network isn't just recognizing a word it has learned," says McClelland. "It's synthesizing a new response. It can give you the correct past tense of completely new words. Even when given a made-up word such as *grok,* for example, the network will produce *grokked.* It learns from experience and *generalizes.*"

Rumelhart and McClelland's network is a very simple model, and not intended to be a complete representation of how children acquire their language ability. But it does provide a demonstration that sophisticated tasks can be performed by a neural network that learns to generalize from examples rather than using logical rules. Perhaps the ability to generalize from experience, not logical ability, is what makes our brains so good at understanding speech and recognizing faces. The ability to draw on previous experiences to respond to new situations makes our brain, like a neural network, an *insight* machine, while the serial rule-and-symbol processing of a conventional computer makes it an ideal *logic* machine. "Connectionism not only accounts for our weaknesses in doing logic problems and the like," says McClelland, "but it also accounts for our abilities to be much better at thinking than anything else is—including machines."

Such irrationality may also provide the glue that holds societies together, such as our compassion for people we don't know or may never see again and our unwillingness to do something because "it doesn't seem right," even though it may be a strictly logical move for other reasons. Our irrationality may also account for economists' problems in making forecasts; economists often assume that people will behave logically when making financial decisions. Ultimately, our irrationality may account even for the fact that we are able to fall in love, despite what reason might say about the fate of many such relationships.

In fact, our irrationality, along with its good and bad consequences, is the trait that distinguishes us from every other creature— and machine—on the planet. It, and not our ability to do logic, is what makes us human. Our blundering, irrational brain creates our hatred and bigotry, envy and paranoia, pride and greed. But it is also behind our ability to enjoy music, forge a sense of justice, believe in things we can't see, and empathize with strangers. It's the source of hope, love, the movies of Charlie Chaplin, and our ability to get out of bed in the morning with the knowledge of our mortality.

CHAPTER THREE

THE GREAT DIVIDE

Brain Science versus Mind Science

It requires a very unusual mind to make an analysis of the
obvious.
—Alfred North Whitehead

Metaphysics may be, after all, only the art of being sure
of something that is not so, and logic only the art of going
wrong with confidence.
—Joseph Wood Krutch

Patricia Churchland's small office, huddled within the white walls
of the library building at the University of California's San Diego
campus, isn't much different from what you'd expect of a philosopher's lair. Except for one thing: Nestled among the stacks of
papers and shelves loaded with books is something that probably
wouldn't be found in any other philosopher's room. In a far
corner, a little rumpled, hangs a white laboratory coat.

"I would like to think that I had deep intellectual reasons for
going to medical school," says Churchland. "But part of it was the
same thing that motivated people I didn't know at the time,
connectionists like Jim Anderson and Terry Sejnowski and Dave
Rumelhart. It seemed to me that the standard model of cognition
used in philosophy, artificial intelligence, and cognitive psychology—
the model that talks about symbol manipulation and rules—*had* to
be wrong. I thought there ought to be answers that you could
get from studying the brain, answers that would *really* explain
things that cognitive psychologists were trying to explain, such as
how we see, how we understand, how we figure things out, what
reasoning is, and so forth. I thought there should be a *physical*
reason for it."

In a way, philosophy can be thought of as the study of the obvious. Most of us don't really think much about whether our minds are distinct from our brains, whether we have a free will or what it means to be conscious or rational. Nor do we dwell too much on epistemology—the nature and limits of what we know and how we organize that knowledge—or ponder what kind of "language" our brain thinks in or what kind of representations of the outside world it uses when it thinks.

While most of us take these aspects of the mind for granted, considering how they might be duplicated in a *machine* seems far from obvious. How can a computer have "consciousness" or "free will"? At a basic level, however, the brain is also a machine. A wet, messy, living one, to be sure, but its billions of neurons act together in an enormously complex mechanism that produces thoughts, consciousness, and questions about thoughts and consciousness.

The World Within

Most philosophers prefer to think about what they consider the essence of thought—the high-level concepts of high-level cognition—and leave unraveling the mind's biological machinery to neuroscientists. That is why Churchland's foray into brain science is so unusual. Her journey began while she and her husband Paul, also a philosopher, were professors at the University of Manitoba in Winnepeg, Canada. There she began to question whether a philosopher could investigate the qualities of the mind without paying attention to the mechanics of the brain. One philosopher who motivated her was Harvard University's W.V. Quine. "Quine made the claim that there isn't a real distinction between philosophy and science," says Churchland. "He said that if we really wanted to make strides in epistemology, we couldn't just sit back in an armchair and search for some sort of 'fundamental laws of thought'—we had to be in the swim with the rest of science."

What are the limits of what we can know, and how is that knowledge organized? How do we represent the outside world to our inner selves? We know that there is a world out there, and we know that that world is somehow represented in our minds. We even know that sometimes our senses gather information from the world that is different from what the world actually is. When we look at a house, for example, we know that it has four walls, even though we can't see all of them.

How do we represent that knowledge in our minds? How do we take the sensory data from seeing a house and infer that there are other walls? We seem to get along perfectly well without contemplating how we do this, but imagine trying to make a machine draw these inferences. How do you organize information so that a machine knows which concepts it needs to perform a task? What are the relevant concepts?

Traditional philosophers, including such giants as René Descartes and Immanuel Kant, have approached these epistemological questions as exercises of pure reason. These philosophers search for the *a priori* foundations of human knowledge, uncluttered by concerns about the nature of the brain. In that sense, epistemology is a discipline of reflection, not experimentation, and is considered outside the realm of science.

Quine, however, argued that epistemology shouldn't be an *a priori* discipline, but an *empirical* discipline. While the term *experimental epistemology* might seem oxymoronic, Quine thought that philosophy should incorporate the findings of science and forge them into an overall picture. "Quine said that," says Churchland, "but he didn't actually *do* it. So I figured, well, why don't *I* do it? Then I thought, how do I start?"

One place to start, thought Churchland, was with the idea of mental representations. Representations are what your mind uses to think about the outside world. For example, most people who study the mind believe that when you are asked to name your favorite flavor of ice cream, you have something somewhere in your brain that represents the concept of "ice cream." You also have in your brain representations of the various flavors. Somehow, you mull all those representations over and come up with "pistachio."

For years, philosophers and cognitive scientists have considered the mind's representations to be symbols. Like the coins in your pocket, symbols have no intrinsic meaning; they represent whatever those exchanging them agree they should mean. In much the same way that the letter e represents "energy" in the equation $e = mc^2$ in physics, in the traditional cognitive science approach "ice cream" could be represented in your mind as a symbol—*ic,* say. The various flavors might be represented, too: c for chocolate, v for vanilla, p for pistachio, perhaps joined with another symbol, a number—1, 2, or 3—that represents how much you like that flavor. These mental symbols are processed in your mind by the rules of formal logic. To determine which flavor of ice cream is your favorite, your mind scans through the various symbolic representations of ice cream and uses these rules to find which flavor is ranked highest.

In the traditional model of our thought processes, the mind is a system of symbols manipulated by logical rules. These rules and symbols form a "language" of thought, the structure of which mirrors the structure of the language that we use to speak to each other. Thus, at some deep level, a French pastry maker, a Tibetan monk, and a Hollywood movie star all use the same basic mental structures to think, structures that can be defined in terms of rules and symbols.

The implications of this traditional model are far-reaching, for if the mind operates by manipulating universal symbols with a universal set of logical rules, then the mind can be studied independently of the brain. Researchers can concentrate on examining the nature of those rules and symbols, and ignore the biological complexities of brain tissue. Because such a system of universal symbols and rules can also be run as a computer program, machines can be used to mimic the processes that presumably go on in our minds. Not only could scientists test their models of mental operations on a computer but they could also incorporate such models in a machine to give it intelligence. To most scientists, this model of the mind is very appealing, and for the last thirty years the main thrust of cognitive science has been to discover the mind's "program."

But according to Churchland, such an approach is a little like that of the man in the old joke who loses his keys in a dark alleyway in the middle of the night, but searches for them around a nearby lamppost because the light is better. Looking for the symbols and rules that make up the "program of the mind" exempts researchers from crawling around in the tangled, complex mess of the brain's neurons. But even though the "light" is better at the symbol-and-rule level of understanding cognition, the mind may not work that way at all, and the marginal successes of the approach might keep scientists from looking for the *real* processes behind our cognitive powers.

A New Engine of Thought

When her husband, Paul, first suggested that the mind's representations weren't symbols in logical sentences, Churchland thought he was crazy. But at Manitoba they had the freedom to be unorthodox, and the two philosophers began seriously to explore the question, urging each other on and becoming more radical. "We began to pursue the idea that maybe what we call thinking and reasoning and understanding and so forth is nothing at all like using a bunch of symbols and going through a little syllogism," she says. "Occasionally, of course, you do go through a deductive argument, but most of the time it's the brain going *mulchity mulchity mulchity*, doing something else entirely. Most of our reasoning and information processing—the smart-dumb business that gets you from input to output—probably bears no resemblance at all to the rules of logic. Of course, the next question was, If the mind's information processing isn't sentence processing, what *is* it?"

Philosophers typically try to answer this question, observes Churchland, by sitting in an armchair and asking, "Holy cow, what could thinking be like?" But Churchland decided to look instead for clues to the workings of the mind in the workings of the brain. "I figured that if I studied neuroscience and found out as much as I could about the brain, then maybe *that* would give an answer as to what a model of the mind should really look like."

Churchland took a sabbatical from teaching and began taking regular classes on brain science with medical students at the university. The researchers in the neuroscience department were very receptive and gave her, as they gave all medical students, a Tupperware pot with a human brain in it. Churchland went on rounds with neurologists at the hospital and soon was helping neuroscientists perform experiments in the laboratory.

Churchland's adventure into neuroscience culminated in her book *Neurophilosophy*, which attempts to combine neuroscience and philosophy toward a new approach in understanding the mind and brain. Now at the University of California at San Diego, the Churchlands are championing a new role for philosophers in neuroscience. he is also exploring a new role for brain science—and connectionism—in the study of philosophy. "It's still too early to tell," she says. "But my hunch is that we are in for some real surprises."

That a philosopher like Churchland had to go back to school to learn neuroscience—and that such a project would be considered a radical move—shows how far the science of the mind has split from the science of the brain. To most nonscientists, this division might seem a little odd, yet neuroscientists are infamous for their unwillingness to consider theoretical models of how groups of neurons interact. And for the most part, cognitive scientists, philosophers, and computer scientists couldn't care less about how neurons work. As one prominent artificial intelligence researcher snapped in response when questioned about neurons, "I work on the *mind,* not on the brain."

Perhaps part of the reason for the split between research on the mind and on the brain is the sheer enormity of the task. Neuroscientists have their hands full just trying to discover how a *single* neuron works; figuring out how billions of them work together is a task reserved for the distant future. Similarly, the realm of human cognition is so vast that it makes sense for cognitive scientists to start with a more general understanding of what thinking is—and later apply it specifically to the human brain. After all, say traditional cognitive scientists, a Martian encountering a computer for the first time would learn very little about how the machine works by tearing it apart and looking at the circuitry.

In the classic image of the scientific struggle to understand the mind and brain, neuroscientists and cognitive scientists are seen as two teams of diggers, trying to bore a tunnel through a huge mountain by starting from opposite sides. As Sigmund Freud wrote in his *Origins of Psychoanalysis,* "Let the biologists go as far as they can and let us go as far as we can. Some day the two will meet."

The Brain-Mind Split

The banishment of the brain from the study of the mind, and vice versa, began long ago with the work of philosophers who studied logic. "It's our fault originally," explains Paul Churchland. "The old philosophers of science and the logicians were doing AI and cognitive science long before those fields existed anywhere else. But they were doing it without computers; they did it with paper and pencils."

For some two thousand years the formalism developed by the Greek philosopher Aristotle was considered the only form of logic. Based on syllogisms similar to those outlined in the previous chapter, Aristotle's logic was conducted in ordinary language. Premises such as *All men are mortal* and *Socrates is a man* led to a conclusion—*Therefore, Socrates is mortal.* That conclusion could in turn be combined with others to build new conclusions.

But as philosophers built their theorems with greater and greater complexity, the imprecision of ordinary language became increasingly problematic. The seventeenth-century German mathematician Gottfried Wilhelm von Leibniz became frustrated with the vagueness of using language in conventional logic. In the same way that the philosopher and mathematician René Descartes had earlier applied the mathematical formulations of algebra to geometry, creating analytic geometry, Leibniz dreamed of replacing Aristotelian logic with a new logic based on mathematics. His goal was to produce a symbolic language so precise that logic would be reduced to mathematical equations. "If a controversy arises," he wrote, "the discussion between two philosophers need be no more heated than that between two calculators. They only have to take up their quills, sit down before their abacuses, and say: Let us calculate!"

Though Leibniz never came close to devising such a system, his work was continued nearly two centuries later by the English mathematician George Boole and later the German logician Gottlob Frege. These researchers helped develop the language of symbolic logic. To get around the ambiguities of conventional languages, Boole and Frege used symbols. The symbols were put together according to logical operations such as addition and multiplication. These were, in Boole's term, the "laws of thought." Most important, Boole recognized that the truth or falsity of these logic statements could be represented by a 1, meaning the statement was true, or a 0, meaning the statement was false. Reasoning, in other words, could be reduced to determining the truth or falsity of a series of logical propositions.

At the turn of the twentieth century in Cambridge, England, philosophers Alfred North Whitehead and Bertrand Russell took the work in symbolic logic theory a step further. They proposed that the mathematics of natural numbers and Boole's and Frege's laws of symbolic logic were one and the same system. In a monumental work, *Principia Mathematica,* they defined mathematics in terms of logical propositions and showed how by applying logic to statements that were universally agreed upon as true, "axioms," a "formal system" of mathematics could be devised.

This new logic of formal systems enabled philosophers to start at the basement level of truth statements and build a majestic cathedral of theoretical proofs. Even more exciting to philosophers was the notion that they could use formal logic to systematize not just mathematics but *all* of science. Thus began the dominant philosophical school of the early twentieth century, *logical empiricism.* Started by Whitehead and Russell, and carried on by the Austrian philosopher Rudolf Carnap and others, the basic tenet of this philosophical school was that science could be understood in terms of the basic laws and symbols of logic. Russell wrote that he had no doubt that "by these methods, many ancient problems are completely soluble. . . . Take such questions as: What is number? What are space and time? What is mind, and what is matter?"

The goal of logical empiricism was to discover the "syntax" of science, the symbols and rules of a logic-based language through which all knowledge could be systematized. Today logical empiricism still provides the philosophical foundation for the "language-of-thought" model of the mind, as well as linguistics, computer science, and cognitive science.

About the time that Russell and Whitehead were devising their formal systems, psychologists were beginning to banish the concept of "mind" from their experiments. Previously, psychologists had followed Descartes, who held that individuals were privy to the workings of their minds, and relied on subjects to give introspective comments about the mental operations that took place during cognitive experiments.

But some psychologists began to question how much insight we have about our mental processes as we think. Their doubts were bolstered by Freud's suggestion that much of our behavior is governed by processes in our subconscious. Psychologists had also begun to experiment with subjects who had little introspective ability, such as children, the insane, and animals. The famed Russian scientist Ivan Pavlov showed that some types of behavior required no invoking of an internal, purposeful mind at all; they were simply a result of a *reflex* action. In one famous experiment, he showed that dogs could be trained to salivate in response to a ringing bell.

Psychologists began clamoring for a science of mind that was as objective as the physical sciences. Behaviorists like B.F. Skinner argued that psychology must be limited to testable theories that examined the relationship between the outside world and the subject's reaction to that world. Behavior, in other words, should be understood solely as a reflex to an external stimulus. The concept of an internal mind that exhibited intentions and purposes was considered too difficult to test experimentally and therefore unsuitable as part of a "hard" science. As the founder of behaviorism, John B. Watson, wrote in the early 1920s, "Psychology as the behaviorist views it is a purely objective science. Introspection forms no essential part of its method, nor is the scientific value of its data dependent upon the readiness with which they lend themselves to interpretation in terms of consciousness. The behaviorist, in his efforts to get a unitary scheme of animal response, recognizes no dividing line between man and brute." With that, the behaviorists banished the mind from their laboratories. They had a lock on psychology until the 1950s.

A Mind in a Machine

That lock was broken, curiously, not by minds but machines. During World War II, psychologists were asked to help design machines that were more compatible with their human operators. Engineers, on the other hand, were trying to design machines that had human-like abilities. At the heart of many of these humanlike machines were automatic control systems called servomechanisms. When a big gun on a battleship had to be pivoted toward its proper aiming position, a servomechanism would measure the difference between where the gun was and where it was supposed to be, and move the gun until that difference was zero. Engineers were also trying to develop radar scanners that were capable of distinguishing between a bona fide signal, indicating the presence of an airplane, and a signal that was random noise.

To engineers, these automated machines seemed to be "processing" information. Psychologists, still rooted in the behaviorist tradition of considering the mind devoid of intentions and desires, suddenly found themselves surrounded by engineers who were talking about big guns that "wanted" to be somewhere, as if the motion were goal-directed and purposeful, and about radar machines that were trying to "understand" the signals they received.

Psychologists were not allowed to use these terms to describe human behavior, yet engineers were using them to talk about machines! The new development of the automatic control of humanlike machines found a voice in Norbert Wiener, of the Massachusetts Institute of Technology, who in 1948 published a landmark work, *Cybernetics*. Wiener proposed a new science that studied motor control and information processing in both animals and machines. Wiener thought of these systems as using feedback and control to reach an equilibrium with their environment. Most important, Weiner suggested that these processes existed—and could be studied—as entities distinct from any particular machine performing them.

Twelve years earlier, Claude Shannon, then a graduate student at MIT, wrote what is probably the most influential master's thesis of the century. Entitled *A Symbolic Analysis of Relay and Switching Circuits,* Shannon's thesis showed that the symbols and operations of Boole's logic system could be expressed by electronic circuits in a machine. The truth or falsity of a particular proposition could be represented as an "on" or "off" state of an electronic switch.

About the same time, a British mathematician, Alan Turing, was setting out a revolutionary idea. He showed that any conceivable computational task—doing formal logic, for example—could be accomplished on a simple "universal" machine. Turing's machine was theoretical; it had an infinitely long tape like a ticker tape, segmented into squares, and a scanner that could read whether or not a particular square contained a slash mark. Using a prescribed set of instructions, the machine could move the tape one square to the right or one square to the left, erase the slash, or print a slash.

Turing's machine was the embodiment of formal logic, a demonstration of how "rationality" could be produced by a mechanical process. As such, Turing's theoretical machine was the fundamental computing device whence all computers were derived. No matter how intricate a computer's machinery, at a basic level, its actions could be represented by a Turing machine, with the presence or the absence of a slash mark indicating the 1's and 0's of digital code. Most important, any computer based on the fundamental mechanics of his machine can mimic *any* other machine that does computation.

This point was not lost on scientists studying the mind. If the Turing machine could imitate any kind of information processor and the brain was in essence a biological information processor, then it really didn't matter whether the architecture of the brain was considered in studying cognition. Since all computers are fundamentally alike, researchers could ignore the nuts and bolts of the brain's machinery and instead concentrate on finding the *program* that ran that machinery. Evidence for this assumption came from neuroscience as well. In 1943, the neurophysiologist Warren McCulloch and mathematician Walter Pitts wrote a paper called "A Logical Calculus of the Ideas Immanent in Nervous Activity" suggesting that the neurons in the brain worked, in essence, like on/off switches; in other words, the brain was a Turing machine.

Their paper proved an inspiration for the early computer designers. John von Neumann, for example, wrote that "following W. Pitts and W.S. McCulloch, we ignore the more complicated aspects of neuron functioning. . . . It is easily seen, that these simplified neuron functions can be imitated by telegraph relays or by vacuum tubes." Drawing from the work of Turing, Shannon, and Weiner, von Neumann showed how electronic components could actually be combined into a Turing type of computer run by a software program. The powerful number-crunching machines that had been developed to break codes during World War II soon became powerful symbol crunchers, too.

This burst of research into mechanical information processing came to a head in 1955, when Herbert Simon is said to have told his class at the Carnegie Institute of Technology: "Over Christmas, Allen Newell and I invented a thinking machine." The next year, Simon and Newell's computer program—"Logic Theorist," run on a computer called Johnniac, after von Neumann—produced the first computer proof of a theorem. The theorem was one of those that Whitehead and Russell had worked out by hand years before in their *Principia Mathematica*. Simon and Newell's Logic Theorist went on to prove thirty-eight more theorems in the *Principia,* and even produced one proof that was more straightforward than that of the philosophers. The *Journal of Symbolic Logic* wouldn't publish an article on the proof, however, because Simon submitted the article as "coauthored" by Logic Theorist.

Simon and Newell thought that Logic Theorist and their subsequent effort, the General Problem Solver, were more than clever ways to solve logic problems. They put their achievement squarely in line with the study of human cognition. After all, they reasoned, Logic Theorist solved problems in a humanlike way, working backwards, substituting variables, and using a simple type of reasoning. They even experimented with impairing Logic Theorist's abilities by removing some of its knowledge and comparing its performance to their research on humans' solving of problems.

While Newell and Simon thought that their approach would yield insights into how the mind worked, they did not suggest that their machine imitated what is actually going on among the neurons in the brain while thinking. "Our theory is a theory of the information processes involved in problem solving," they wrote later, "and not a theory of neural or electronic mechanisms for problem solving." They suggested that examining the workings of the mind at this symbolic, information-processing level might be more fruitful than trying to understand thinking in terms of the physiology of neurons. If the brain were a Turing-like thinking machine, then might not other Turing-like machines give us insights into the symbols and rules that guide thought? Once the rules and symbols used by the mind were discovered, neuroscientists could determine how neurons in the brain produced them.

As machine scientists began pursuing this approach to studying the mechanisms of the internal mind, psychologists were still employing the behaviorist concept of a "reflex" mind in their experiments. But the machine scientists, with help from philosophers and linguists, overthrew the behaviorists' stronghold on psychology, and the new scientific discipline of *cognitive science* was born. In cognitive science, the mind was seen as the software program that ran on the hardware of the brain, a program whose rules and symbols could be used to govern computers as well. The digital computer served cognitive science both as a model for the mind and (following Turing's assertion that all computers are fundamentally alike) a research tool on which hypotheses about the mind could be tested.

For the last three decades, cognitive scientists have followed this line of inquiry. Ironically, however, as this new model of the mind was being born in the 1960s, the philosophical movement that during the 1920s had laid much of the theoretical groundwork for it—logical empiricism—was being abandoned by philosophers. The dream of systematizing all of science through formal systems had been shown to be impossible by philosophers such as Quine, Kurt Godel, and Ludwig Wittgenstein, and the enthusiasm for the formalist approach had faded. Despite its failure as a philosophical doctrine, logical empiricism continued to have an influence in the world of cognitive science. Howard Gardener writes, in his *The Mind's New Science: A History of the Cognitive Revolution*:

> A major ingredient in ongoing work in the cognitive sciences has been cast in the image of logical empiricism: that is, the vision of a syntax—a set of symbols and the rules for their concatenation—that underlie the operations of the mind. Thus when Noam Chomsky posits the basic operations of a grammar, when Richard Montague examines the logic of semantics, when Allen Newell and Herbert Simon simulate human reasoning on a computer, or when Jerome Bruner and George Miller seek to decipher the rules of classification, or "chunking," they are trying to decipher a logic—perhaps *the* logic—of the mind. This vision comes through even more clearly in the writings of Jerry Fodor, who explicitly searches for a "language of thought" . . . Thus, a model that proved inadequate for scientific enterprise as a whole still motivates research in circumscribed cognitive domains.

Why does logical empiricism, abandoned by philosophers years ago, continue to drive much of the study of the mind today? "You have to realize how appealing it was to consider the mind as a formal system," says Paul Churchland. "There are good reasons for being sucked in. People said, 'The brain's a bowl of porridge— forget it. Let's look at thinking in terms of this universal thinking machine.' " Modeling the mind on a computer allowed researchers to feel comfortable ignoring the complex intricacies of brain tissue. It also allowed researchers to use computers, which were rapidly growing in speed and size.

As cognitive scientists became more and more enraptured with building models of the mind on digital computers, they became less and less interested in looking at the brain itself for clues about the mind's operation. They also moved farther and farther away from considering whether their theoretical models of thought could be carried out by the brain. That the brain has very little in common with a digital computer was never really considered. "If there *were* a distinction between software and hardware in the brain, then the approach was a good one," says Paul Churchland. "But there isn't this clear distinction, so the basic assumption of the whole approach is screwy."

The realization that the operation of the brain has a profound impact on the operation of the mind has been slow in coming to cognitive scientists. As Gardener points out,

> The initial intoxication with cognitive science was based on a shrewd hunch: that human thought would turn out to resemble in significant respects the operations of the computer, and particularly the electronic serial digital computer which was becoming widespread in the middle of the century. It is still too early to say to what extent human thought processes are computational in this sense. Still, if I read the signs right, one of the chief results of the last few decades has been to call into question the extent to which higher human thought processes—those which we might consider most distinctively human—can be adequately approached in terms of this computational model. . . . The kind of systematic, logical, rational view of human cognition that pervaded the early literature of cognitive science does not adequately describe much of human thought and behavior. Cognitive science can still go on, but the question arises about whether one ought to remain on the lookout for more veridical models of human thought.

The inability of artificial intelligence to account for human cognition doesn't mean that the attempt to produce intelligent computers has been a failure. AI researchers have produced some ingenious and very useful programs that in some cases resemble the way we think. But the initial goal of trying to mimic intelligent behavior on computers in order to gain insights into how our minds work has been given up. "Modern computers can do fantastic and marvelous things, and it would break my heart if anybody ever took mine away from me," says Paul Churchland. "They are the most wonderful invention of the last two centuries. *But they don't operate the way that we do.*"

The Changing of the Guard

Like most revolutions, scientific upheavals depend on something to revolt against. The revolution is usually accomplished more quietly in science than in politics, at least outside the confines of a laboratory, but it's often a bruising, agitated process, nevertheless, one that is repeated over and over as old models and theories are discarded and new ones take their place.

In his *The Structure of Scientific Revolutions,* science historian Thomas S. Kuhn suggested that the evolution of science occurs through a process whereby one *paradigm* is replaced by another. A paradigm is a general, overall view of a particular aspect of the world, a fundamental framework in which scientists can formulate questions. For example, when modern astronomers look at the heavens, they conduct their experiments within the paradigm that the universe is expanding. When you go for a checkup, doctors no longer conduct their examinations using the centuries-old paradigm of the body's four cardinal *humors*—blood, phlegm, choler, and black bile (melancholy).

As scientists conduct their studies, they produce evidence that helps refine the paradigm, but sometimes they produce evidence that seems to contradict the paradigm. Eventually, the accumulation of conflicting data is too great, generating a period of crisis when new hypotheses vie for attention. As the dust settles, a new paradigm emerges, and the process starts all over again.

In a general way, the study of the mind has followed this pattern: The early psychologists' introspective approach was overthrown by the behaviorists, whose theories were overthrown by rule-based cognitive science, computers, and artificial intelligence. Is the study of the mind headed for another paradigm shift? According to Paul Churchland, some of the signs are there. "The degeneration of the old research program is now so protracted, so widely acknowledged, and so disappointing that a lot of people are losing faith. The tide is shifting, and now people who study the mind are becoming more interested in neurobiology. The connectionist paradigm is opening the door in an unexpected direction into a space where no one can yet see the limits. People are getting excited."

Given the history of shifting fashions in mind science, Patricia Churchland advocates tempering all the excitement with a dose of caution. "There are probably a zillion different ways of computing, for example, what goes into our visual system," she says. "But if you don't stick close to the neurobiology, you're wasting your time. You can look at the history of cognitive science and say that many people wasted a huge amount of time putting together flow diagrams of the way they thought the mind worked, when they had no idea why their diagram would be more plausible than ninety-nine zillion other ideas."

A complete understanding of the brain and mind will come only if neuroscientists and psychologists support each other's actions; brain scientists need a theory of the mind to guide them in planning experiments on the brain, and cognitive scientists need to know whether their models of cognition are biologically plausible. Churchland thinks that an appropriate image of brain and mind sciences working together is not two tunnelers starting from opposite ends of a mountain, but two climbers struggling out of a narrow crevice, with their backs against each other and their feet on either wall.

Philosophers will also play an important part in the endeavor. Bridging the gulf between the electronic rumblings of nerves and the academic ramblings of metaphysics will depend on cooperation between the differing disciplines of neuroscience, psychology, and computer science. Philosophers can contribute by providing scientists with an overall vision of the workings of the brain and mind. That overall vision may incorporate a new model of the mind, based on connectionist principles, because neural networks have the potential to bring the study of the brain back into the study of the mind.

A New Concept of Mind

The results of that union may have a profound effect on philosophy. "I think it will change our whole concept of how we think about ourselves," says Patricia Churchland. "A neurobiological

understanding of the mind will cause us to rethink our old pre-
sumptions about who we are: what it is to have a self, what it is to
have a soul, what it is to be responsible, to think, to introspect,
and to have a free will. It may have tremendous implications for
morals, too. That's something that other philosophers get very
nervous about. They ask, 'What about responsibility, praise, and
blame?. All these things will be changed.' They're right, all those
things will have to change. I don't know what the result is going
to look like, but they *will* get changed."

Most philosophers believe, for example, that we exercise free
will by examining an array of possibilities before us. We consider
these possibilities, rationally weighing the pros and cons of each,
and then make a choice. But as pointed out in the previous chapter, our decision-making processes often bear little resemblance to
logical reasoning. "Our decision making is much more complicated and messy and sophisticated—and powerful—than logic,"
says Churchland. "Our decision making may turn out to be much
more like the way neural networks function: The neurons in the
network interact with each other, and the system as a whole
evolves to an answer. Then, introspectively, we say to ourselves:
'I've decided.' "

If neural networks alter how we view rationality, then our
notions of responsibility may have to be reconfigured to incorporate that new model of how we think. "If your concept of responsibility is tied to your concept of rationality," says Churchland,
"and your concept of rationality is tied to the old 'symbol and
logic' model, then your concept of responsibility is in trouble."
Neural networks suggest that when we make decisions, we often
don't use deductive reasoning. Since our society's notion of responsibility centers on an individual making rational choices, a
radically new model of how we make those choices might result in
a new way of viewing our responsibility for our actions. "In many
cases no one cares whether a choice is *really* rational or not," says
Churchland. "But in the case of someone on trial for doing something horrendous, we as a society might care quite a bit."

So much of the fabric of our society is woven by our ideas of
praise and blame, responsibility and morality. In many ways, those
concepts arise out of our traditional notions of a rational mind. We
are reluctant to punish children and others for actions over which
we think they have no rational control, and we often blame others

for acting "irrationally" by following their intuition or instincts. These problems arise especially in areas such as the public's perception of risk, which is often regarded by experts as irrational. Yet neural networks, with their shift away from a purely logical approach towards an emphasis on bringing many aspects of a problem to bear on a solution, may in fact help us create a new sense of ourselves that is broader and, in some ways, more humane.

The questions a neural network model of the mind raises about concepts of rationality and responsibility suggest that as we better understand the inner workings of our mind, society will need philosophers more than ever. "There are philosophers who think what I'm doing is really horrible because it disturbs the traditional way of thinking about humans," says Churchland. "I imagine people said the same thing to Copernicus. But I don't think the prospect of change is discouraging, I think it's hopeful and very exciting."

Because the new model of the mind is still in its infancy, it will be a long time before such a revised view of ourselves takes final shape. But replacing the old view of the mind as a "logic machine" with a new model that incorporates the complex dynamics of interacting neurons may lead to a view of ourselves that is much more humanistic. "There seems to be a long-standing tradition that science is the enemy to be fought," she says. "But I absolutely reject the idea that science is pitted against humanism. It's true that there are people who abuse science in the name of politics, but that's not *science*. If there is an enemy to humanism, it is superstition and prejudice."

With their insistence that understanding the brain is essential to understanding cognition, philosophers are once again laying a foundation for a new model of the mind. Unlike their predecessors nearly a century ago, however, the new philosophers may have an influence that extends beyond cognitive science and philosophy. Connectionism is also reaching those researchers who have long labored in the dark alleyways of tangled brain tissue. For neuroscientists, the new model of the mind may be the first step to switching on the light.

CHAPTER FOUR

WETWARE

The Anatomy of Memory

Memory can restore to life everything except smells,
although nothing revives the past so completely as a smell that
was once associated with it.
　　　　　　　　　　　　　—Vladimir Nabokov

Gary Lynch bounds into the room, excitedly thumbing a copy of
Newsweek. The issue's cover story is about the brain, and at the
beginning of the story, in large bold type, is a quote: "Memory is
the black hole at the center of neurobiology." Attached to the
quote is Lynch's name. He waves the magazine. "I've got to get
some more of these. They're going to stop printing them, and I'm
not going to have any to give my mom."

A man with a round face and bright eyes, Lynch shows an
almost childlike delight in being the center of attention—where his
brash outspokenness often lands him. Lynch is a neuroscientist, a
researcher who puts on his galoshes and slogs around in the wet,
slimy mush that you use to read this sentence. While Patricia
Churchland might wonder whether it is possible to make sense of
the complexities of the mind while sitting in the sterile confines
of a library, neuroscientists like Lynch have the opposite problem of
trying to understand the mind while engulfed in the complex biology
of brain tissue. Neuroscientists work in cognitive quicksand; the
more they struggle for an understanding of the mind, the more they
become bogged down in the physiological details of the brain.

What's worse, there seems to be no way to get free. Every
time they try to step away from the brain's biology onto the more
pristine land of theory and hypothesis, neuroscientists carry a little
of the wet stuff with them. It is hard to make grand theories and
simplified models when each brain cell is oozing exceptions. Because of these complexities, neuroscientists are notorious for staying

away from the larger issues of just how the nerve cells they study so intimately actually join together to produce the mind. Most of them have their hands full figuring out how just *one* of the hundred billion or so neurons in the brain works.

The traditional resistance of most neuroscientists to large-scale theorizing means that they will probably be the last researchers to embrace the neural network model of the mind. In some ways, their wariness is justified. Through the years, neuroscientists have politely ignored countless researchers who boldly announced that their new theory was going to explain how the brain works. Behaviorism, American structuralism, German Gestalt, ethology, the Freudians, artificial intelligence—all were going to provide the ultimate explanation of the mind, and all faded from fashion. "It's a terrible thing to say," says Lynch, "but I think artificial intelligence never said anything of any interest to people in brain science, except for those who were mathematically oriented and wanted to do equations."

Constructing elegant theories of how the mind works without consulting the facts of how the brain works is risky. Time and again neuroscientists have pulled the rug out from under psychological theories with physiological data. For example, years ago, there were several psychological theories of how we perceived colors, but they all vanished in an instant when neuroscientists demonstrated that our perception of color was simply a chemical reaction. Says Lynch, "The assumption has always been that we are going to vanish the psychologists and the AI guys because we are going to find the chemicals that make the memory and the circuit that makes the behavior. While that's the standard chauvinistic attitude in neuroscience, I don't subscribe to it. We will *not* explain things on wet data alone—there will have to be theory that sits at the top of neurobiology and psychology and incorporates the wet data."

That theory, says Lynch, may well be connectionism, because neural networks do attempt to understand the mind through a knowledge of the brain. As news of neural network research spreads throughout the brain-mind community, more and more neuroscientists are beginning to pay attention, and a few brain researchers are even taking a few tentative steps into the land of modeling and theory themselves. Alan Gelperin, of Princeton University and AT&T Bell Laboratories, for example, is working with physicist John Hopfield, of the California Institute of Technology, on using a neural network to simulate learning behavior in the nervous system.

Other neuroscientists such as Terry Sejnowski and Christof Koch are helping to create a new discipline called computational neuroscience, where researchers explore how groups of neurons link together to process information. Koch and computer scientist Tomaso Poggio, for example, are studying the computational properties of nerve cells in the eye's retina—how the cells detect motion when an object is traveling across the visual field. Jim Bower, of the California Institute of Technology; Richard Thompson, of the University of Southern California; and Edmund Rolls, of the University of Oxford, have also begun to venture from studying the action of a single neuron to examining the activities of *groups* of neurons.

Talking to Lynch is both exciting and dangerous, a little like driving a sports car on a road hugging a mountainside. The scientific research behind Lynch's speculations is solid, but he likes to push it a little, hoping to get near the edge for a breathtaking view. Lynch has a reputation among his colleagues as one of the few who are willing to make bold leaps into hypotheses. His actions sometimes raise a few eyebrows in the neuroscience community, perhaps the most conservative of brain-mind sciences. But Lynch's intelligence and ebullient personality make him a central figure among a small group of neuroscientists who are challenging their colleagues to look beyond the strict confines of their data on single neurons to the question of how the brain works as a whole. "I was recently at a scientific meeting where the purpose was to see if we could go from the level of the neuron all the way out to the cognitive realm of the mind," he says. "It got *completely* out of control. That's what launched me down the road to get more and more into neural network theory."

Probing the Mind's Machinery

Lynch's assaults on the brain are conducted in his laboratory at the University of California at Irvine, which looks like what you'd expect from a scientist who trudges among tangled neurons: dishes, wires, tubes, recording machines, and a patina of disorder. "This lab spans the neurobiology world," says Lynch, strolling through the cluttered rooms and waving his arms. "The largest single group of people working here are biochemists, and most of the work being done here is on neurons. The place I do the most work is in the trailer that I've got out back. The university ran out of office space, so they gave me a trailer. It's quiet, and I can smoke my cigars without bothering anybody."

We walk into a room containing a large microscope flanked by wires and an oscilloscope. Under focus is a tiny dish. A loud hum, sounding a little like a transformer at a power station, emanates from one of the machines and fills the room.

Lynch turns to a graduate student sitting nearby.

"You got anything in there?"

"Yes."

"Is it alive?"

"Just barely. It's been in there about ten hours."

In the dish is the hippocampus of a rat. The hippocampus is a tiny region of the brain that appears to be involved in encoding some kinds of memory.

"You can pop the hippocampus out of a rat brain any time you want," says Lynch. "We study it just like it was an electrical circuit. We just reach in there, pull that baby out by its pins, put it in a dish, and do just like electrical engineers working on a computer chip. We now know the microanatomy—the wiring diagram that goes in there—and in some cases we know it with exquisite precision. Take a look. There are a lot of people in this world that have never seen a hippocampus."

It doesn't look like much—gray mush with tiny black dots floating in it and a tiny wire coming out of it. "We use that wire to stimulate the tissue with electropulses. That makes the neurons operate the way they normally do inside the brain—they propagate the pulse through their circuitry. You can't actually see the circuit lines. But we know where they all are."

"You see all these little black knobs? That's where the inputs from other neurons connect. When you realize that this is just *one* tiny branch of the neuron, you begin to get a feeling for how many of these knobs must be here. There are about fifty thousand of these places on each cell. *Terrifying.*"

Lynch points to the tiny wire in the tissue. "Getting the microelectrode into a single neuron is great sport, a little like fish-

ing. Everybody loves to do it. You take a glass probe, which has a
tip that's about a micron wide, and you move it down through the
tissue until you come up to a cell membrane. You use a high
current to charge the membrane, and it opens up. That hum you
hear is the noise of the electrode as it is going down through the
tissue. When it gets inside a cell, you'll hear it scream. It's the
eeriest sound. There's this quiet hum and then '*BURRRRRRRP-
PPPPP!!!*' Then you know you're inside a single neuron."

Recording from single neurons has been the dominant mode of
brain research for nearly three decades. Neuroscientists David Hubel
and Torsten Wiesel used the technique to demonstrate that individual neurons in a monkey's brain respond not simply to the intensity of light, but to specific patterns of visual stimuli such as sharp
edges and lines. Hubel and Wiesel recorded the electrical activity of
neurons in a monkey's visual cortex, part of the brain that helps
process vision. They discovered that the neurons were arranged in
small groups, each responsible for analyzing a tiny spot on the
eye's retina. They also found that some neurons were most active
when the monkey looked at a dark stripe oriented at a specific
angle. Other neurons responded most to stripes at other angles,
bright lines at different angles, or the edges of shadows. Hubel and
Wiesel's research, for which they won a Nobel prize, had a large
influence on other neuroscientists. "Their research," says Terry
Sejnowski, "launched a hundred thousand microelectrodes."

Microelectrode studies and research into the psychological
impairments that result from brain damage due to strokes or other
head injuries have given neurobiologists a good idea of where
some parts of our mind reside in our brain. Make each of your
hands into a loose fist, and put them side by side. That's about the
size of your brain. Each fist corresponds roughly to one of the
brain's two hemispheres. Where your wrists join is the brain stem,
the place where the brain meets your spinal column. Where your
fingers curl under are some of the structures that regulate the
bodily functions that keep you alive—the thalamus, the hippocampus, and the pituitary gland.

Covering your hands and fingers, wrinkled and folded like a
hand towel stuffed into a coconut, is the cerebral cortex, your
thinking machine. Because our thinking ability distinguishes us
from other species, we humans tend to be preoccupied with this
eighth-inch-thick sheet of neurons. But the brain itself isn't preoccu-

pied with thinking. The brain is designed primarily for doing other things: moving the body, sensing the environment, staying alive. Of course, since most animals do that too, most of us don't find those brain functions very interesting.

For all of our attention to our thinking powers, however, we know very little about the cerebral cortex. Neuroscientists have located some parts that are responsible for the body's motor control and other parts that do sensory processing. But much of the cortex isn't responsible for any body function; it's used for reasoning, speaking, and remembering. While no one knows how this happens, a lot has been discovered about the way the foot soldiers of the brain, the neurons, work.

Gary Lynch has a poster-size photograph in his office that shows a greenish structure against a black background. The green thing looks like a hairbrush whose bristles have run amok or a time-lapse photograph of the biggest fireworks on the Fourth of July. Green fibers spurt everywhere in all directions; you can barely see the spine where they attach. (See the Figure below.)

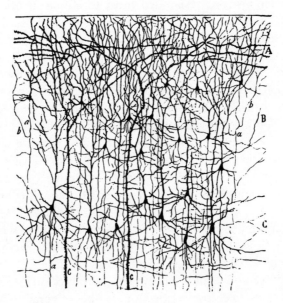

This drawing made by pioneer neuroscientist Ramón y Cajal shows the dense interconnectedness of neurons.

"That's a neuron," Lynch says with a chuckle. "I like to show it to connectionists, because they always say, *'What the hell?* So *that's* what they look like. I thought that neurons were like little lines.' Well, we draw them that way, but they aren't. This neuron has fifty to a hundred *thousand* inputs."

Neuroscientists are a little wary of the connectionists' new model of the mind because they know that neurons are far more complex than the simple units used in neural networks. There are many types of neurons in the brain and different types of neurons in different animals. But in fact, neurons are more alike than they are different. The most significant difference between your brain and that of a hamster is that your brain has many more neurons and that they are organized and connected together much differently.

A basic picture of a neuron would include a nucleus, the central body of the nerve, and short fibers extending around the nucleus, called dendrites. (See the Figure on the following page.) Most neurons also have a single longer branch, called an axon, that has smaller branches at its end. A neuron usually receives information from other neurons via its dendrites and sends messages out to other neurons via its axon. (That's not always the case; in some neurons, dendrites also send messages, axons send messages to other axons, or dendrites influence the actions of one another.)

In most mammalian neurons, these messages are electrochemical; that is, the action of a nerve is powered by electricity and sent by chemicals. This complicated process is different for different kinds of nerves, but in mammals it goes roughly like this: When a neuron fires, an electric pulse travels down its axon at nearly 150 miles an hour. When the pulse reaches the end of the axon, it causes holes to open at the end. Chemicals flow out of the holes and into a tiny gap, called a synapse, that lies between the end of the axon and a dendrite of another neuron. (See the Figure on page 65.)

The chemicals are some of the thirty or so "neurotransmitters" known to exist in the brain. Which chemicals flow out depends on the kind of neuron. The chemicals migrate across the synapse to the membrane of the adjacent nerve's dendrite, attach to receptors there, and a chemical reaction takes place that makes the receiving neuron either a little more charged or a little less charged.

There are thousands of dendrites on a typical neuron, and as more and more reactions take place, the charge inside the neuron builds until it reaches a certain threshold. Then the neuron sends its

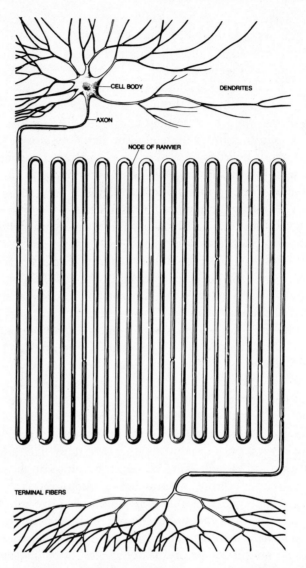

A typical neuron of a mammalian brain has a large cell body crowned by spreading dendrites at the top and a long axon (unrealistically folded to fit in this illustration) that ends in wide branching. A neuron's signal travels down the axon to the end of these branches and triggers the release of chemicals. The chemical signals travel across a tiny gap, called a synapse, between the axon of one nerve and the dendrites of another.

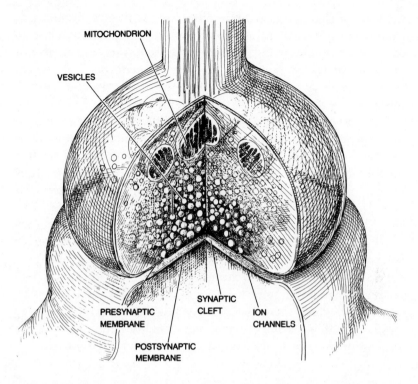

MITOCHONDRION

VESICLES

SYNAPTIC
CLEFT

PRESYNAPTIC
MEMBRANE

ION
CHANNELS

POSTSYNAPTIC
MEMBRANE

The synapse is the area where the signal from one neuron is transmitted to another. Chemicals released by the axon of one neuron travel across the 200 nanometer gap to the receiving dendrite. These chemicals in turn influence the electrical charge of the receiving neuron, making it more or less likely to fire its own signal.

own charge down its axon to other neurons. Through the elaborate interplay of millions of these chemical reactions, a winning chess move is chosen, a thigh muscle contracts to catapult a basketball player high into the air, or a fond memory emerges from the sweet smell of a summer day.

As discussed in the previous chapter, neuroscientists McCulloch and Pitts helped launch the computer revolution with their suggestion that neurons switch on or off, but in fact, the action of neurons is often gradual. A neuron's pulse, once started, keeps going down the axon to the end, but this "all or nothing" characteristic is the only thing it shares with its electrical counterparts in computers.

While the neuron's electronic pulse is quick, its final effect—

the flow of chemical neurotransmitters across the synapse—can be a much more gradual process. An axon may release various amounts of transmitter; a receiving dendrite might have varying amounts of receptor; the transmitter itself may have different chemical properties and react at different rates. And the whole process may be mitigated by the action of various enzymes.

A nerve is not like a simple relay circuit; whether it fires or not depends on the complex interplay of many inputs, both inhibitory and excitatory, from all the neurons connected to it. If a neuron doesn't get enough excitatory input from the neurons connected to it, or gets too many neurotransmitters that inhibit neural action, it will do nothing.

Most important, the communication lines between neurons can be changed. The ability of synapses to be made stronger or weaker is critical to neural network models of the mind: It is generally believed that by changing the strengths of synapses in our brain, we learn, store memories, and alter our behavior. The process by which these changes occur is still a mystery. It may involve anything from growing new synapses, to changing the amount of receptor or transmitter, to the action of enzymes.

Wiring the Brain

In 1949, Canadian neurophysiologist Donald Hebb proposed, in his *Organization of Behavior,* a simple model of how learning might happen: "When an axon of cell A is near enough to excite a cell B, and repeatedly or persistently takes part in firing it, some growth process or metabolic change takes place in one or both cells such that A's efficiency as one of the cells firing B is increased." In other words, if one neuron sends a lot of signals that excite another neuron, the synapse between the two neurons is strengthened. The more active the two neurons are, the stronger the connection between them grows; thus, with every new experience, your brain slightly rewires its physical structure.

Hebb's hypothesis was speculative, but neuroscientists have recently found evidence suggesting that he was right. Eric Kandel of Columbia University, for example, has shown that in the tiny sea-slug *aplysia,* the connections between the animal's neurons grew stronger as it learned to associate a food it disliked with the presence of a beam of light.

It is unclear whether studies from an animal as simple as *aplysia* can be applied to the human brain, but other research has found a similar learning effect on neuron communication in mammals. Working with tissue from a rat's brain, researchers have found that if two connected neurons are stimulated at the same time, the amount of signal passing from one neuron to the other can double. It is uncertain whether the change in synaptic strength, known as long-term potentiation or LTP, is permanent, because the tests are done on nerve tissue that can be kept alive for only a few weeks.

Though neuroscientists are still struggling to understand the chemical processes involved in altering the connections between neurons, they have come to realize that carving out these pathways plays a crucial role in wiring the brain to produce the mind. You have about a hundred billion neurons in your brain, about ten billion of which are in your cortex. You lose about one thousand to ten thousand neurons each day after you reach forty. If that's alarming, consider that you were born with virtually all the neurons you will ever have, and during the first years of your life, as much as 85 percent of them were deliberately killed off.

This mass destruction was necessary, however, if you were to become a thinking being, for while some of your neurons were dying, the surviving neurons were growing, wiring the circuitry of your mind. Their growth is prodigious; when you were born, your brain was about a quarter of the size it is now.

The brain's neurons lay down their circuitry of connections during the first few years of infancy. While some of this wiring is genetically programmed, much of it is not. *Aplysia,* for example, is the object of much neural study because its twenty thousand nerves are large and well defined. Each and every synaptic connection among those nerve cells is specified by its genes. "You could almost go as far as to say that each individual neuron in *aplysia* has its own *color,*" says neuroscientist Richard Thompson, of the University of Southern California (USC). "Each individual neuron is identifiable and reproducible, and every nerve in *aplysia* is identical to the nerves in other *aplysia.*"

But in the human brain, the number of synaptic connections is millions of times greater than the number of human genes. It is virtually impossible for genes to encode the entire wiring diagram of a human brain—or even those of less brainy animals. One particular species of tropical fish, for example, reproduces by cloning; that is, each fish makes an identical copy of itself. Though a cloned pair of fish have exactly the same genes, no two of them have the same synaptic connection pattern in their brains.

If the genes don't direct the wiring of all the synapses, what does? It appears that for many connections between neurons the architect is experience. In one study by a team of scientists led by William Greenough, of the University of Illinois, a group of rats was raised in a large "penthouse" cage filled with toys, and another group of rats was raised in sparse cages. The brains of the rats raised in the rich environment developed neurons with 20 percent more branching and contacts to other neurons than those in the brains of the rats raised in the barren environment. The rich environment apparently made the brains more active, and they responded by growing more complex connections.

The effect of the brain's experiences on neurons' wiring is strikingly evident in the human visual system. When a baby is born, both eyes are connected to all the neurons in the brain's visual cortex. As the newborn develops, however, each eye competes with the other for neural pathways. As one eye takes over the pathways in a particular area in the brain, the other eye's neural circuitry is shut down. Usually, each eye winds up with roughly half of the neural pathways—the visual cortex of a normal adult brain is striped with alternating patterns of right-eye–left-eye connections.

This neural wiring is orchestrated by stimuli to the eyes themselves. Recent studies suggest that if a neuron is being used, it secretes a substance that affects nearby cells responsible for the neuron's nourishment. These cells, in turn, produce a chemical that appears to preserve the neuron from destruction. If the neuron does not get that substance, it dies.

If for some reason one eye receives no stimuli during this crucial time of development, the other eye will take over *all* the available neural circuitry in the cortex. For example, if an eye is temporarily closed because of an injury, the brain circuitry for that eye may disappear. The eye will still function normally, but the brain won't see anything through it. Visual stimulus is so important for neural development that a kitten with one eye closed for one day will have permanently impaired vision in that eye. The same is true for humans. Neuroscientists say that parents should make sure that their newborn babies don't have one eye closed for very long; it's better to have both of them closed.

How the brain of a growing fetus might arrange its wiring to perform visual tasks is demonstrated by a neural net simulation created by Ralph Linsker at IBM's Watson Research Lab in Yorktown, New York. Linsker's network is arranged in a hierarchy of several layers of neurons arranged in a sheet, with groups of neurons in one sheet connected to various individual neurons in the sheet above it. When information comes into the first layer, the network modifies connections between neurons that are active at the same time, so that their signals are preferentially sent on to the next layer.

Linsker starts his model with random inputs, but as the signals travel upward through the layers, the neurons adapt their communication channels, organizing themselves into "specialists" for different types of visual inputs. At the top layer of the network, for example, specific neurons respond to bars of light in specific orientations. These cells are quite similar to the pattern-specific neurons found in the monkey's brain by Wiesel and Hubel. Linsker's work may help explain why the brain is able to develop these areas that respond to specific types of visual scenes while still in the womb where it is getting no visual signals.

While the brain undergoes the most changes during childhood, there is evidence that neurons in some areas retain their ability to modify their wiring into adulthood. Michael Merzenich, of the University of California at San Francisco, mapped out the different areas in a monkey's brain where groups of neurons process different fingers' sense of touch. Then Merzenich trained the monkey to use one finger predominantly in a task that earned it a food reward. When Merzenich remapped the touch-processing neurons in the monkey's brain, he found that the group responsible for processing signals from the much-used finger had expanded 600 percent.

Researchers Leif Finkel and Gerald M. Edelman, of Rockefeller University, simulated this type of neural rearrangement in a neural network. The artificial neurons in their model arrange their connections according to a theory proposed by Edelman called

Neural Darwinism. In a way analogous to Darwin's natural selection, in which animals will compete with each other to put more of their offspring into the next generation, groups of neurons in Edelman and Finkel's network compete with one another for neural pathways.

Edelman's network initially has only random connections between its neurons, but as the network is exposed to incoming information, the neurons organize themselves into groups specialized for different kinds of information processing. The organization process begins when information—touch stimuli from a finger, for instance—first comes into the network. The information activates some groups of neurons more than others, and this high level of activity causes the connections among the group of excited neurons to be reinforced. As more and more similar patterns come through the network, the connections among the activated group of neurons become stronger and stronger, and eventually the group becomes specialized for processing that one finger's sense of touch.

When new experiences arise, the neurons rearrange their connections. Edelman and Finkel mimicked Merzenich's experiment with monkeys, allowing their network to organize itself to process touch from several fingers, then giving one finger a large input. Like the neurons in the monkey's brain, the neurons in the artificial network rearranged themselves in response to the increased input, expanding the number of pathways they used to process the suddenly increased amount of information.

A neural network's ability to arrange its connections in response to experience makes it an ideal tool for modeling cognitive processes in the brain. In a study by brain scientists Richard Anderson, of the Massachusetts Institute of Technology, and David Zisper, of the University of California at San Diego, the researchers were trying to determine how the brain is able to locate precisely an object in front of it from nerve signals coming from the eyes. The computation involved in such a task is extremely complex. If you move your head from side to side as you read this sentence, for example, your eyes will shift position as you turn your head, moving the images of the words to a different part of your retina. Yet the brain is able to compute continually where in space those words are despite the fact that the two things it relies on to do the task—the position of your eyes and where the image is on your retina—is constantly changing.

To find out how the brain might perform such a feat, Anderson and Zisper trained a neural network by giving it signals recorded from neurons controlling the muscles that move the eyes and those in the retina. After training, the neural network could judge the position of an object in front of its "eyes" by itself. When the researchers then took readings from the various artificial neurons in the network, they found that one set of neurons produced signals very similar to those found in the area of the brain that does this task, strongly suggesting that they had correctly modeled one tiny—but extremely sophisticated—aspect of the brain's visual processing ability.

Making Memories
(or What Memories Are Made Of)

A neural network's ability to change its own connections may also give researchers insights into how the brain stores information as memories. Cognitive scientists have discovered that we have several distinct types of memories. One of the first clues to our multifaceted memory came from the case of a man known as H.M., part of whose brain was removed to alleviate his severe epileptic seizures.

The surgery made H.M.'s seizures less severe, but it also had the tragic consequence of permanently altering his ability to remember. If you met H.M., you might not notice anything unusual about him in the course of a conversation. He has an above-average IQ and can talk about a variety of topics. But ten minutes after he left the room, H.M. would have no recollection of you or that he'd ever talked to you. He knows his name and retains all the memories he had before his surgery, but can't encode new memories of facts.

However, H.M. can still encode new memories of *procedures*. He can learn, for example, to hit a slice serve in tennis. Shown the proper way during a lesson, he will improve with practice and retain the knowledge like anyone else. "But imagine being H.M.'s tennis instructor," says USC's Thompson. "He will learn the actual skill—he just can't remember what it is called, anything you said about it, or, for that matter, who you are." The ability to remember procedures also applies to nonmotor skills. If you give

H.M. the same puzzle to do every day, for example, he won't remember having done the puzzle, but he will solve it quicker as the days go by. Somehow, H.M. has learned the procedure for solving the puzzle without remembering having solved it before.

Cases like H.M.'s have led cognitive scientists to divide our memories into different parts: procedural memory and declarative memory. Some researchers divide declarative memory further into fact memory and episodic memory. When you encode a procedural memory, you are learning how to do something; if you solve twenty puzzles with the same format, for instance, you would probably do the twentieth puzzle much faster than you did the first one. The same is true for learning to play tennis and checkers or learning to read writing reversed in a mirror. The more you practice the procedure, the better you perform, and the better you remember it. Fact memory is what you use when you are asked, "What is the capital of Italy?" If you are asked about your vacation in Italy last summer, however, you aren't just recalling facts about capitals and countries; you are telling a story, and the tales you relate about Rome would be episodic memories.

These different types of memory have different properties. Procedural memories are learned through practice and are thought to be long-lasting; hence the cliche about knowing how to ride a bicycle forever, once you've learned how to do it. They are also recalled best when you are actually performing the action. For example, try *thinking* through the motions of tying a shoelace, without moving your hands.

Fact memories can be long-lasting, but they can be hard to get at. In one study, people were asked to memorize a list of obscure words and their definitions. The subjects were then given a list words that contained words that had been previously memorized and some new words. When the subjects were stumped for a definition—but said it was "on the tip of their tongue"—95 percent of the time it was a word they had memorized before.

Some cognitive scientists suspect that in episodic memories, events are not so much recalled as reconstituted. You may remember some parts of the memory at some times and other parts of the memory at other times. Much of your recall of your vacation in Italy, for example, may depend on whether the person you are talking to is giving you clues. For example, your listener might ask what kinds of foods you ate, whether you went to the mountains, or whether you visited a vineyard. The more detailed the questions, the more you may remember.

In fact, some researchers think that other peoples' suggestions may be inadvertently incorporated into the memory of an event. This can be a problem for eyewitnesses at a trial. "Suppose you take ten people and show them a film of a crime," says Gary Lynch. "Then you ask those ten people to tell you what they saw. And suppose you ask the first person, 'Did you see them get away in that green Mustang?' When another person comes up and is told, 'Tell me exactly what you saw,' the person may say, 'I saw them come out and get into a green Mustang,' even though there actually was no green Mustang. This is controversial, but some psychologists believe that the person actually has the green Mustang *in memory*. He is picking up the fact and incorporating it into his memory."

Where memories are stored in the brain is still a mystery. Some studies suggest that memories are stored in specific places; people undergoing brain surgery occasionally speak of entire scenes and images welling up in response to stimulating a tiny part of their brain tissue. One patient who suffered brain damage from a stroke appears to have his linguistic abilities intact, but he cannot remember the names of common fruits and vegetables.

Yet other researchers have found that memories' locations seem more spread out. Years ago, the psychologist Karl Lashley taught a group of rats to run a maze, then destroyed various parts of their brains to discover where the memory of the maze was stored. Lashley found that he could not remove the memory of how to run the maze by removing any one part of the brain, though the animal's overall performance deteriorated. Many patients who lose some cognitive functions as the result of a stroke eventually regain these functions. Scientists now believe that while general areas of the brain perform specific tasks, the activity is spread over many neurons. Just as there are different types of memories, there may be different ways they are stored in the brain.

Most people dwell on the failures rather than the successes of their memory, but their brains are better at remembering than they realize. Consider how many words you know, (it's upwards of seventy-five thousand), or how many people you can recognize by sight. In one experiment that tested the limits of the old saying "I never forget a face," Ralph Haber at Cornell University showed students more than two thousand slides of people and scenes, one after another. The next day, Haber showed his subjects pairs of slides—one slide seen before and a similar slide not seen before. The people correctly picked the slide they had seen before 90 percent of the time.

On the other hand, some aspects of our memory are quite limited. Look at the number 6826442. Now look away and see if you can repeat it. Now try the same thing with 501639753245. Can you look away and remember it? If you can, you have a better short-term memory than most people. Psychologist George Miller showed that people usually can hold only about seven items in their short-term memory at one time. Of course, it is possible to memorize a long number if you repeat it to yourself over and over. For example, many people know their Social Security number by heart. But that requires passing what is held in short-term memory on to the more permanent storage of long-term memory. Nobody knows how your brain does that.

While it is true that memories are not always 100 percent accurate, the brain's inability to store mountains of detail actually may be a critical part of its remarkable abilities to make generalizations, match patterns, and intuit. After all, having *too* good a memory can cause problems. The famed Russian psychologist Aleksandr Luria wrote of one man who had a memory so vast that he made a living putting on demonstrations of memorizing large groups of numbers and words. But the mnemonist complained that he was miserable because he could not *forget* things. He also had terrible difficulty remembering faces because, since people change their looks from day to day, he could not match a slightly different face with the old one he remembered.

The brain's memory is not as accurate as that of a typical computer, nor is it stored in the same way. In a computer, one set of circuits is responsible for storing the memory, and other circuits process the information; there is evidence that in the brain, memory and processing are merged in one system. The same circuits that *process* memories are also responsible for *storing* them.

A Brain Circuit for Smell

To find how these memories might be stored, Lynch is studying the brain's circuitry for processing odors. "Smell is the only sensory system that has *direct* access to the brain regions that encode memory," he says. "Vision, auditory, touch, and the motor system all go through these regions, too, but the connections are extremely tortuous—tracing their wiring is like trying to outline the connections between all the telephones in New York City."

The brain's wiring for smell begins with the nose, the only place where your brain is directly exposed to the outside world; it literally dangles its neurons out into the air. The neurons in the nose connect to neurons in a smell-processing area of the brain called the olfactory bulb, which in turn connects to the olfactory cortex, which links to the hippocampus. Your hippocampus, therefore, is only a couple of neural connections from the outside. The simplicity of this circuitry may explain why a scent can conjure up sights, sounds, and feelings but doesn't work in reverse. "If I say the word *rose,* it's trivial to think of an image of one," says Lynch. "But you can't just think of a rose and recall the smell of it. There's no feedback from the other sensory structures. That's why it's so hard for us to recall smells."

By selectively cutting the pathways of the smell circuit of a rat, Lynch has been able to outline the kinds of neural circuitry that produce different kinds of memory. Cutting one part of the circuit makes the animal no longer able to associate smells with procedures. Cutting a different pathway prevents the animal from encoding fact memories about smell. Lynch suggests that these experiments indicate that the brain uses one type of neural network for fact memory and a different type for procedural memory.

While working on the brain's smell circuitry, Lynch found that he had unknowingly stepped into the land of connectionism. At a presentation of his research at a scientific meeting several years ago, Lynch met Geoff Hinton, a computer expert now at the University of Toronto. "Hinton looked at my stuff and said, 'Wait a minute, you're saying something that is very familiar to me.' The brain circuits I was studying were extremely similar to neural network designs for memory. I discovered, to my surprise, that I was a connectionist."

Lynch and a colleague, computer scientist Richard Granger, have begun to test their theoretical network models of the brain's smell circuitry in a computer. In one experiment, their five-hundred-neuron network was presented with two groups of simulated odors, each containing variations of a more general pattern representing "cheese" and "flower." Each simulated odor was represented by a pattern of neural activity: A cheese would stimulate one group of neurons, and a flower would stimulate a different group.

Lynch found that at first the network would respond with a unique pattern of activity for each odor. As the network was presented with more and more examples of similar odors, however, the network began to evolve. Neurons that were repeatedly activated by similar odors became stronger and stronger, eventually dampening those neurons that were less active. Gradually, these highly active neurons came to represent the general category that the group of odors belonged in.

After being exposed to a half-dozen examples of similar smells, the network would respond with the same overall pattern of activation on the first sniff whenever any smell was presented, identifying the general group to which the smell belonged. On subsequent sniffs, however, something unexpected happened: The old pattern disappeared, and a new pattern of activation arose unique to each smell. "We're thrilled with it," says Granger. "With the first sniff, it recognizes the overall pattern and says, 'It's a cheese.' With the next sniffs, it distinguishes the pattern and says, 'It's Jarlsberg.'" Lynch has found some support for his model in experiments with rats. He found, for example, that over the course of sampling an odor, specific neurons in a rat's brain, like those in his neural network, are active in the presence of some smells but not others.

Lynch's neural network lacks the cognitive sophistication of many other connectionists' neural nets, partly because it stays very close to the structure of real brain tissue. "My work comes out of neuroanatomy," he says. "Everything begins with the wiring in the brain and proceeds from there. It's as if somebody gave me a wiring diagram of a radio or a computer and said, 'How does this work?'"

Starting from the biology of the neurons themselves and working his way up to theory, however, allows Lynch to check the actions of his model against lab experiments using real brain tissue, something only a few other connectionists can do. "We can talk to post-docs who do anatomical work and say, 'See if it really works this way.' Of course, they always come back and say, 'No, it doesn't.' That's true—we're not very close now, to be completely honest. But that's okay, because we're getting closer and closer." His attention to the real details of the brain also allows the neuroscientist in Lynch to sleep a little easier at night. "There was a connectionist at a meeting recently who said that 'all the relevant biology that we need to make neural networks has been discovered.' My response is I don't know of *any* real biology they've put in a network. But there are some connectionists—the better ones like Sejnowski and Hinton—who are aware of that, and they are working on it."

Lynch's willingness to go beyond his brain-circuitry data to theoretical models of how they work may cause some neuroscientists to cringe, but his work is a source of comfort to connectionists who worry about whether their models really capture the essence of brain-style computation. "The circuits in my networks are not hypothetical," he says. "This is the real thing—this is *brain.*"

One intriguing biological detail, says Lynch, is that the connections in the neural net circuits for smell are essentially random; that is, there appears to be no overall structure built into them. That's very different from the kind of circuitry the brain uses to process vision, for example. Much of the neural circuitry for vision is what neuroscientists call *topographic;* that is, inputs from a particular area of your eyes' retinas are mapped directly onto a particular area of your visual cortex. If someone holds up a finger in front of you, for instance, a neuroscientist can tell you where on your brain's visual cortex that finger is being seen. Furthermore, it's the same place on everybody's visual cortex.

But when you sniff an odor, says Lynch, there's no way of knowing where in your olfactory bulb you are sampling that smell. "We have no idea where it is, because the response among the neurons is distributed. When you smell a rose, it's perfectly possible that the neurons in your olfactory bulb are doing something completely different from mine." Because there is no topographic "map" of the olfactory world, smell is cognitive from its very beginnings. "Smell isn't spatial or temporal. It doesn't exist in a dimensional world," he says. "It's like pure thought."

The Evolution of Mind

This smell circuit may have led to the evolution of the other cognitive areas of our brains as well, suggests Lynch. Smell is the sense that characterizes mammals. Among vertebrates that live on land, only mammals have highly evolved smell systems. Fossil evidence indicates that when the mammalian brain evolved its large cortex, the olfactory system also expanded, and in simple mammals, the cortex is still largely occupied with olfaction. When mammals began to develop the neural circuitry for smell, says Lynch, evolution discovered a type of neural network useful for sophisticated thinking. "My argument is that the initial expansion of cortex in mammals may have come as a result of this greater dependency on smell," he says. "Now from there, you can really begin to develop some extraordinary hypotheses."

When the dinosaurs disappeared about 65 million years ago, says Lynch, mammals suddenly found themselves expanding into new environments that called for better vision. The type of neural circuitry that worked so well for the olfactory systems began to be used for vision as well. "When mammals suddenly found themselves in a much richer visual world," says Lynch, "*vision discovered connectionism*—quite literally." Our visual system may be a marriage of traditional AI and connectionism, says Lynch. Part of our visual system works algorithmically; specific neurons are tuned to detect stripes oriented one way or another. Another part of the system may use a connectionist neural network to assemble the various pieces of information into a coherent whole. "What emerges," says Lynch, "is something that really happens in vision—top down and bottom up working together."

Lynch admits that this particular theory about our visual system's evolution is fairly speculative. "But I think that there's enough data out there to make you wonder if it's not the case," he says. "Maybe we have so much trouble understanding how the brain does vision because we are so fascinated with this topographic map, and we ignore the nontopographic components of vision."

A neural net created by Terry Sejnowsi and his colleague Sidney Lehky provides further proof that it may be incorrect to infer a neural system's functioning as a whole from the actions of individual neurons. Sejnowski and Lehky trained a neural net to detect how various objects are curved by the way their surfaces are shaded. After forty thousand trials with two thousand different shading patterns, the network learned to make the assessment of curvature by itself with about 90 percent accuracy. The surprising part of the study came when Sejnowski and Lehky analyzed the activity of individual neurons as they made a decision. They found that while each neuron responded only weakly to the shaded figures, it responded with the *most activity* when the network's input neurons were given not shaded figures, but *bars of light or dark*.

Sejnowski's network raises questions about the conclusion many brain researchers have drawn from the famed studies of Hubel and Wiesel. When Hubel and Wiesel found that individual neurons responded with the most activity when a monkey looked at dark lines of various orientations, many researchers concluded that these neurons helped the brain detect the edges of objects in visual scenes. The neurons in Sejnowski and Lehky's network respond just like the "edge-detecting" neurons found by Hubel and Wiesel, but they were trained on nothing but shaded figures that had no edges! "It just demonstrates the trouble you run into when you try to understand the brain from a strict 'bottom-up' perspective," says Sejnowski. "You can't simply infer the *overall* function of part of the brain—'Oh, it's a boundary detector'—from looking at how an *individual* neuron responds."

The work of brain researchers such as Sejnowski and Lynch suggests that as more and more details are discovered about how our brains process sensory inputs, store information, and retrieve memories, scientists will have to shift away from studying how neurons act individually toward studying how groups of neurons interact as an integrated system. This is a change that scientists from many other disciplines are beginning to make, and neuroscientists are realizing that as in the weather, the nation's economy, and other extremely complex systems governed by interactions among many simple elements, the activity of the brain as a whole may be quite different from the sum of its individual parts.

CHAPTER FIVE

WHAT MAKES A BUNCH OF NEURONS SO SMART?

The Science of Complexity

And new Philosophy calls all in doubt,
The element of fire is quite put out,
The Sun lost, and th'earth, an no man's wit
Can well direct him where to look for it.
And freely men confess that this world's spent,
When in the Planets and the Firmament,
They seek so many new, then they see that this
Is crumbled out again to his Atomies
'Tis all in Pieces, all coherence gone.
—John Donne, *An Anatomy of the World*

Chaos is the score upon which reality is written.
—Henry Miller, *Tropic of Cancer*

Like a neural network, the connectionist movement derives its power through the interaction of diverse elements: Terry Sejnowski and Gary Lynch provide a neuroscientific foundation for the new model of the mind; Patricia Churchland adds a philosophical perspective; and Jay McClelland, David Rumelhart, and Geoff Hinton contribute psychological applications and new neural network designs.

John Hopfield, a trim, nattily dressed man with a wry smile, is a relative stranger to the wet world of neurons and the dry musings of psychologists and philosophers. Yet his contribution to connectionism is crucial, for he provides the glue that holds together the ragged pieces of experimental data. To connectionism,

Hopfield, a physicist at the California Institute of Technology, brings mathematical theory.

In a paper published in the *Proceedings of the National Academy* in 1982, Hopfield demonstrated mathematically the process by which a large group of simple interacting neurons join to process information. The brilliance and insight of Hopfield's paper, "Neural Networks and Physical Systems with Emergent Collective Computational Abilities," provided a guiding light for connectionists around the world. "We always knew that neural nets worked, but Hopfield showed *why* they work," says Brown University's Jim Anderson. "That was really important, because it gave us legitimacy."

The Mathematics of Mind

Hopfield has always tried to explore uncharted areas of scientific research. During the 1960s, he worked in solid-state physics, investigating the properties of silicon and other materials used in electronic devices. When those devices became commonplace in computers, Hopfield looked for new scientific realms to explore. In the late 1970s, he began meeting with a small group of scientists who gathered twice a year at MIT to discuss neuroscience. Researchers from all areas of brain science were in the group; Hopfield was invited along as the resident outsider, someone who knew enough about physics and chemistry to contribute to the discussions.

As the meetings progressed over the years, Hopfield gradually absorbed an education and, always on the lookout for new scientific horizons, set out to find a problem to solve in the wet, sloppy realm of the brains. He began exploring the mathematical properties of networks of interacting neurons. The neural network he studied bore little resemblance to real neural circuits, but its simplicity allowed him to use his mathematical background. "I was trying to create something that had the essence of neurobiology," Hopfield says. "My network is rich enough to be interesting, yet simple enough so that there are some mathematical handles on it. It's almost as if I were taking a step backwards, simplifying this neural network model in order to be able to do honest things mathematically."

Hopfield's insight into neural networks was as simple as it was brilliant: The process by which a group of interacting neurons eventually come to a decision can be thought of as a physical system whose energy is decreasing. Consider a ball rolling down a hill, the smell of a pot roast spreading throughout a house as it cooks, or a stretched rubber band snapping back. The ball, the heated air molecules from the roast, and the rubber band can be thought of as physical systems that are changing from one state to another. In their initial state, each system can be characterized as containing a large amount of energy; you could use the energy from a stretched rubber band, for example, to shoot a wad of paper across the room. In their later state—the ball at the bottom of the hill, the cooled air molecules distributed throughout the house, the rubber band in its relaxed state—the systems have less energy. Over time, each system as a whole evolves from a high-energy state to a low-energy state.

Hopfield showed mathematically that a network of interacting neurons could be thought of in the same way. At the point of receiving an input, the network is in a high-energy state. As the neurons talk among themselves and produce a decision, the system evolves toward a lower-energy state. Hopfield's neural net "thinks," in other words, not in the logical, step-by-step manner of a typical computer, but more like a New England town meeting, where an initially heated debate gradually settles into a decision.

Scientists have long been aware that simple interactions among large simple elements can produce complex behavior. Weather is the result of simple interactions between billions of individual air molecules; the interplay of millions of single cells in the blood gives rise to the body's immune system; billions of neurons link together to form the brain. While the interactions between any two individual elements in each of these systems are as simple as a molecular collision, a chemical reaction, or the firing of an electronic pulse, their cumulative effect can be far more complex. From these tiny interactions arise hurricanes, the body's defense against disease, and the mind.

While the interactions between individual elements in such systems are relatively simple, understanding how those millions of interactions join in a complex system has long been considered beyond the reach of science. Since Galileo first measured the motions of falling balls and swinging pendulums, scientists have concentrated on trying to understand the universe as if it were made up of objects that look and behave like utopian billiard balls rolling on a vast utopian pool table.

A Wrench in the Gears of the Universe

This mechanistic vision of the universe blossomed during the eighteenth century. The universe was seen as a giant clock in which God, the Prime Mover, pushed the biggest gear and everything else followed in concert. Confident scientists asserted that if you knew the state of all the elements of the universe at one time, you could predict what the universe would be like years into the future. The clock worked in reverse, too, in exactly the same way; the same set of conditions that foretold the future could reveal the past.

"Why did the clock almost immediately become the very symbol of world order?" ask Nobel laureates Ilya Prigogine and Isabelle Stengers in *Order out of Chaos.* "In this last question lies perhaps some elements of an answer. A watch is a *contrivance* governed by a rationality that lies outside itself, by a plan that is blindly executed by its inner workings. The clock world is a metaphor suggestive of God the Watchmaker, the rational master of a robotlike nature. . . . Scientists were in the process of discovering the secret of the 'great machine of the universe.' "

Despite a great deal of intellectual development since—and many scientific discoveries to the contrary—civilization for the most part has not abandoned this idea of a mechanistic, predictable, clocklike universe. Though this model is simplistic, it yielded many scientific discoveries, because in many cases it is possible to reduce phenomena to straightforward laws. Newton and Einstein revealed some of the universe's hidden gears with their theories of gravitation and relativity. Using these theories, astronomers can calculate where each planet in the solar system will be hundreds of years from now and are beginning to understand what the universe was like billions of years ago during the first few seconds of its birth—and how it might end billions of years hence.

But by the beginning of the twentieth century, there were signs that in some ways this approach had reached its limits. Scientists working in the arena of subatomic physics found that, in fact, some things—the very essence of matter, for example—are not predictable. At this level, which is governed by the laws of physics known as quantum mechanics, matter—electrons, photons, and other particles—is still interconnected, but not so that scientists can make absolute predictions about the future.

Other phenomena in nature are chaotic: while they display general predictable patterns, their state at any particular time is unpredictable by scientific equations. Meteorologists know that it will be hot in Miami during July and that it will snow in Alaska during the winter. But because the behavior of the weather system as a whole is extremely sensitive to changes in any of its billions of parts, scientists cannot use the state of the weather at one time to predict with absolute accuracy what the weather will be like on any particular day; in theory, a butterfly flapping its wings in Hawaii could affect the weather in New York three days later.

Such unpredictability may also be a feature of the brain. "It may be that our feelings of original thought and free will are the result of the fact that very small influences can totally change the brain's state from one time to another," says Hopfield. "Like the weather, the state of your brain three days from now may be changed by two words that you hear now."

These kinds of complex systems defy understanding by traditional scientific methodology. Scientists typically try to understand phenomena by focusing on a simple example of it; physicists studied atoms of hydrogen, the simplest element, to get a fundamental understanding of how more-complex atoms of other elements were constructed.

But some complex systems can't be understood by focusing on their parts. "For some phenomena, the essence of the system *is* its complexity," says Carver Mead, one of the founders of computer chip technology and now a pioneer in making neural network computer chips. "That's a new idea in physics." Because these systems are so difficult to understand, most scientists concentrate instead on problems that yield to straightforward analysis. But in fact, most natural phenomena are complex systems, and regarding them as "special cases," the mathematician Stanislaw Ulam once observed, is "like referring to the class of animals that are not elephants as nonelephants."

Over the last decade, however, the exotic worlds of complexity have begun to yield their secrets. One newly emerging scientific discipline, known as *chaos theory*, is demonstrating that the seemingly random behavior of phenomena such as rising smoke from a cigarette, the growth and decline of animal populations, and the weather has an underlying systematicity. The work of Polish-born mathematician Benoit Mandelbrot has brought another corner of the irregular world closer to the realm of scientific understanding. Called *fractal geometry*, Mandelbrot's research concerns objects whose shapes, unlike regular geometrical objects such as planes or spheres, combine both randomness and regularity. Mountains or snowflakes, for example, have a familiar form that everyone recognizes, yet the details of one snowflake are quite different from another. Phenomena as diverse as the clustering of galaxies in the universe, the branching of the human lung, and the coastline of England have been shown to have an underlying fractal geometry.

Miniature Universes or Miniature Worlds

Scientists are also trying to understand the dynamics of complex systems by studying cellular automata—groups of elements that interact with each other according to simple rules, but produce complex, sometimes chaotic, behavior. Studies of cellular automata are almost always carried out in the miniature universes of computers. A researcher starts with a set of elements, applies a set of rules for their interaction, and lets the computer work out how such a population might evolve.

One of the simplest examples of cellular automata is a game called LIFE, invented by the British mathematician John Horton Conroy in 1970. LIFE is played on a large grid resembling a checkerboard. To begin, any number of small counters—checkers, for example—are placed on the squares of the board in any pattern. Then, by playing out a set of simple rules, this "population" of counters is made to grow, reproduce, and evolve through generations.

The rules are simple:

1. The eight squares surrounding any particular square are its "neighbors."
2. Any checker that has two or three neighboring checkers has enough support to survive until the next generation.

3. Any checker that has four or more neighbors dies from overpopulation and is removed.
4. Any checker that has only one checker as a neighbor dies from lack of support.
5. Any empty square that is surrounded by three neighboring checkers—no more or less—has a birth in it, and a checker is placed there in the next generation.

Births and deaths occur simultaneously, so the game proceeds by laying out a pattern of checkers, then using the rules to create the next "generation" of checkers.

A simple pattern of three dots such as this, for example,

will evolve to this pattern in the second generation,

and then disappear in the next generation.

A three-checker pattern like this, however,

will evolve into a block

and stay that way forever.

And a pattern like this

will evolve to this

in the next generation, then back to this,

switching back and forth into eternity.

This simple checkerboard universe can give rise to some amazingly complex populations and behavior. Many patterns continue to evolve for dozens of generations, then disappear or settle into a stable state. Some become oscillators, recreating the same pattern after a fixed number of generations. Some patterns glide across the board indefinitely:

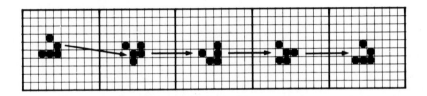

Other, more intricate patterns evolve into what are known as "glider guns"; that is, the pattern goes through a repetitive cycle lasting a number of generations, in the process shooting a glider across the grid each time! Since every new glider adds more checkers to the board, this population grows forever.

The study of cellular automata was pioneered in the early 1950s by John von Neumann. Von Neumann suggested that it was possible to create a self-directed machine, an automaton, that would reproduce itself forever, thereby growing into a huge population of copies of itself. Such machines have been the subject of more than a few science fiction epics, including the first *Star Trek* movie. Von Neumann used a theoretical board a little like that in the game of LIFE, but his rules for how generations evolve were more complicated, and his squares, rather than being either empty or filled with a checker, could have any of twenty-nine possible states. Using such a system, Von Neumann was able to prove that such self-reproducing cellular automata can exist.

With the advent of superfast computers during the last decade, research in cellular automata has blossomed. Norman Packard of the Institute for Advanced Study has demonstrated that by defining each square in a grid as either "ice" or "vapor," and using rules that mimic the mechanics of freezing, he can grow cellular-automata "snowflakes" that look like the real thing. Other researchers insert a dab of probability into their models. Richard Durrett of Cornell University is using cellular automata to simulate the spread of a forest fire. In his model, each cell in the grid represents a tree, and the rules for evolution mimic the chances of fire spreading from one tree to another. Other cellular automata model the growth of a plant population or the trickle of an oil spill through a bed of sand. (See the Figure on the following page.)

While these studies of artificial complex systems are mostly theoretical, there is hope that someday they will lead to an understanding of real phenomena such as the body's immune system— the collection of molecules and cells in our blood that fights harmful viruses, bacteria, and other pathogens and the part of the body ravaged by the AIDS virus. Packard, along with physicist Doyne Farmer and biologist Alan Perelson, both of the Los Alamos National Laboratory in New Mexico, uses strings of 1's and 0's to represent antibodies (the molecules in the blood that recognize harmful invaders) and similar strings of 1's and 0's to mimic the

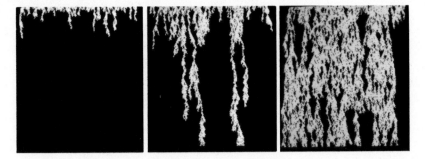

Computer models of complex systems simulate processes that involve many interacting particles, such as this idealized scenario of an oil spill percolating through sand. When the sand is packed tightly and therefore is not very porous, the oil migrates downward only a little before stopping, as in the illustration at left. If the sand is slightly more porous a few trickles of oil continue downward (center illustration), and at higher levels the trickle becomes a downpour (right).

molecular "signature," called an antigen, that an invading pathogen carries. The researchers program various antibodies and antigens into the computer simulation, then let them interact. If an antibody's number string is close to that of an invader's antigen, the antibody binds to the invader, kills it, and produces more copies of itself. The stronger the match between antibody and antigen, the more quickly the antibody interacts and reproduces. Researchers are also using computer simulations to study how complex molecules emerge from a primordial soup of smaller molecules, mimicking the process thought to have given rise to life on Earth. Other scientists are trying to model the turbulent airflow around an airplane wing and how an intricate pattern of markings on a seashell might be created by a simple set of genetic instructions.

Not only do complex systems require a new approach to scientific understanding; they may also cause scientists to redefine what "understanding" is. Traditional science tries to explain phenomena in terms of causes and effects. But the new science of complexity is demonstrating that an explanation such as "x leads to y, which results in z" may be impossible for some systems, because these direct causal relationships don't exist. "We're going to have to broaden our notions of what an explanation is," says Terry Sejnowski. "We will be able to solve these problems, but the solutions won't look like the neat equations that we're used to." A scientist trying to answer the question "What is someone thinking?" may encounter the same problems as the hapless weather forecaster trying to get an exact weather prediction for a particular place and time. Scientists may be able to answer the more general question, "How do people think?" however, in the same way that meteorologists can understand the dynamics of the different inter-actions that lead to clouds, rain, and hurricanes.

Neural networks have the potential to help explain how we think because they have many characteristics of a complex system. In Hopfield's network, for example, each artificial neuron, like the squares in the game of LIFE, can be in one of several states; in more sophisticated neural nets, the neurons can take on a broader range of values. Like the cells in a complex system, the state of any one neuron depends on the states of the other neurons connected to it, and the state of each of those connected neurons depends on the inputs it receives from all the others connected to it, and so on.

The neurons in a typical neural net actually bear little resem-blance to neurons in the brain. Connectionists like to say that their neural networks are "brain-style" devices. The networks behave in a way that resembles—though grossly oversimplifies—the action of neurons. A neuron in a neural net acts somewhat like a pressure valve in a system of water pipes. This "valve" can be set at a certain level; if it gets enough input from other neurons to go beyond that threshold (in the water-system analogy, enough water from other pipes), the pressure will open the valve, and water will be sent along the pipes toward other valves.

Suppose a particular neuron in a network is designed to send a signal when it gets a positive input of 4 or more. If the network gets a signal of 1 from four other neurons connected to it, it will send out its own signal. But the situation could be more complex. The neuron, for example, could be connected to twenty—or two hundred or two thousand—other neurons. Some neurons could be sending a message of +5, others, −3; others could be sending no message at all. The neuron adds all of these inputs, and if the incoming flow from surrounding neurons goes over its threshold, the neuron turns on, sending a message to the neurons to which it is connected. That message might in turn affect the states of other neurons, and so the process goes until the switching stops.

This dynamic interaction among neurons gives neural networks their ability to recall memories and make decisions. This "thinking" ability, however, isn't the result of any one neuron's action. Rather, it *emerges* from the complex interaction of large numbers of individual neurons. "Many of the mysteries of the brain are what in physics we could call *emergent properties*," says Hopfield. "They arise from the interaction of very large numbers of elements. Though they are the consequences of millions of microscopic relationships, emergent properties often seem to take on a life of their own."

But assembling a collection of interacting neurons isn't enough to produce thinking abilities. Neither evolution nor connectionists can throw a group of neurons together and expect them to start suddenly doing algebra or writing symphonies. Emergent properties may arise spontaneously, but our brain required millions of years of evolution. As interconnected groups of neurons displayed various types of information-processing abilities, they were selected, modified, and adapted. "Biology doesn't particularly care where the emergent properties come from," says Hopfield. "It just uses whatever evolution gives it that happens to work. If emergent properties occur in a particular network of neurons, and evolution can use them for computation, that's all the better."

Hopfield's is one of the simplest designs for neural networks. In it, all the connections between neurons are symmetrical; that is, if neuron *A* talks to neuron *B*, then *B* also talks with equal strength to *A*. To help understand the emergent properties of his neural network, Hopfield likes to demonstrate how you can make a neural network out of a thousand high-school students, a thousand desks, a gym, and a few simple electronic parts.

First, sit each of the students at a desk. Then on each desk, put a battery, an on/off switch, and an electric meter that measures the current coming into it on a scale of, say, 1 to 10. Next, take a huge roll of wire and connect each student's battery to every one of the other students' electric meters; put the on/off switch in the circuitry, so that each student can disconnect his or her battery from all the outgoing wires.

The last step is a little complicated. There are electrical components, called resistors, which reduce the amount of current flowing through a wire. By putting a resistor in a student's circuit, the amount of current going from one student's battery to another's meter is diminished: the bigger the resistor, the smaller the current. For the final touch on the student neural net, put resistors of various sizes in the circuits.

The student neural net is complete: You have a gymnasium full of students, each sitting at a desk and capable of flicking a switch that sends a current to all the other students' desks. There are resistors in some wires making the current between some students stronger than that between others. On each desk is a meter that adds all the current coming in.

Now you are ready to put the student neural net in action. First, you wander through the gym, randomly stopping at a few hundred of the students' desks. At some desks, you clamp the switch permanently in the "on" position; at others, you clamp the switch permanently "off." Then you go to the front of the room and announce a final set of instructions: "Each student must stay at his desk, but can go about his business, studying, reading, or whatever. Every once in a while, look at your electric meter. If it reads about 5, then turn your switch on; if it is already on, leave it on. If the meter reads below 5, turn the switch off; if the switch is already off, leave it off. If you are one of the students whose switch is clamped, do nothing."

You give the signal to begin. At first, there is a commotion of switches turning on and off, as one student sees his or her meter rise above 5 and turns the battery switch on, sending current to the other students. That current, in turn, will cause other students to turn their switches on. Other students will see that their meters are below 5 and turn their battery off, decreasing the current flow to the others. The currents flowing in and out of the students' meters will rise and fall.

Hopfield has shown, however, that eventually *all* the students will stop switching their switches. Those desks with meters reading above 5 will have their switches on; those with meters reading below 5 will have their switches turned off. And the network will stay that way.

Finding the Best Solution Among Many

Hopfield has shown mathematically that this student neural net has some remarkable properties. Since there are a thousand students in the neural network, there are a huge number of different configurations the on and off switches can assume—10^{300}. But Hopfield has shown that no matter how the switches are set at the start, after a period of switching commotion, the thousand-student net will settle into one of only a hundred or so final configurations. The secret is the resistors. Since some connections between students contain smaller resistors than others, some students are more influential than others and push the network towards one of a small number of final states.

In Hopfield's mathematical language, the student neural net can be thought of as trying to lower the amount of energy it contains. Hopfield uses the analogy of a landscape laden with hills and valleys (see the Figure below). If you flew in a plane over this landscape and threw a bucket of water out of the window, the

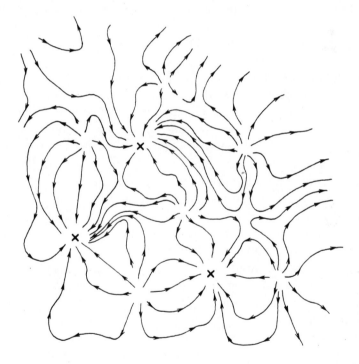

Cal Tech physicist John Hopfield pointed out that a neural network could be thought of as having a "landscape" with valleys and hills. Posing a problem to the network is like pouring water onto the landscape: it flows down the "hills" into a valley that usually represents one of the best answers.

water would hit the ground and flow downhill into the nearest valley. In the same way, says Hopfield, the various resistors in the student neural net give the network "hills" and "valleys." The thousand-student network might have a hundred such valleys of low energy among high-energy places that correspond to hills. When the switches in a network are set at random, it's like throwing water out of the airplane onto the terrain. Like water flowing downhill, the network evolves towards the nearest low-energy state.

This decision-making process makes a neural network very different from a computer in the way it "thinks." A typical computer moves methodically from step to step, passing information along a chain of logical decisions in a process not unlike decision making in a typical business hierarchy. Each worker makes a recommendation to his or her immediate supervisor; the supervisor tallies those opinions, makes a decision, and then feeds it up the line to the next level of supervisors. These people in turn make their decisions and send them along the line. By the time the information has reached the president of the company, it has all come together, and a final decision is made.

But in neural networks—and the brain—information flows back and forth as different elements in the system work together. The workers, or neurons, tell one another how strongly they feel on the issue. As they listen to the discussion, the workers change the strength of how they feel. Gradually the opinions of the workers polarize and the group comes to a decision.

Both types of information processing can be useful; hierarchical structures are perfect for doing high-level mathematics, bookkeeping, and other difficult tasks that involve precise details and exact answers. "Town-meeting" types of structures are more flexible and freewheeling in their information processing and so are less suited to tasks requiring precision. But that freewheeling style may be better at optimization problems where incoming information is often slightly inaccurate or incomplete and a choice must be made from many alternatives.

One such problem is pattern matching, as when the incoming odor of a rose must be paired with one of thousands of smells stored in memory. Another is trying to find the sequence of arm movements that will get a robot's hand from one place to another most quickly. Often an optimization problem will ask for a minimum: the shortest route, the least amount of wire, the closest fit, the nearest match. Neural networks are good at solving such problems because they operate, as Hopfield demonstrated, by minimizing their energy. That lowest-energy state represents an optimal solution.

A typical optimization task is known as the Traveling Salesman Problem, which commonly arises in everyday life, from routing traveling salesmen, to creating airline schedules, to designing microchips—any situation where there are many ways of doing something and the challenge is to find the shortest route. Suppose, for example, that you are a salesman and have to visit ten cities in North America. What's the shortest route you can take to visit them all at least once?

For ten cities there are 181,440 possible routes. While it might be a manageable task to find which of these 181,440 routes are the shortest, as the number of cities to be visitied increases, the number of possible routes skyrockets. For one hundred cities, for example, there are more than 10^{100} routes. Though there are sophisticated programs for solving this problem on digital computers, their basic strategy is simply to measure each route, one by one, and that takes a lot of time.

On Hopfield's neural net, resistors can be put in the connections between the neurons to represent the different distances between the cities and the order in which they can be visited. Within a few millionths of a second, the machine settles into a low-energy stable state that represents one of the shortest routes. In one experiment by Hopfield and his associate at AT&T's Bell Labs, David Tank, a neural network found an answer to a Traveling Salesman Problem a thousand times faster than a conventional computer.

A neural network is able to choose among millions of possible answers quickly because it doesn't consider each answer one by one. Instead, the network takes a somewhat different approach. Each of the many possible solutions to a problem can be thought of as a true-or-false proposition: It is *true* that a route is the shortest, or it is *false*. When a conventional digital computer solves the problem, it goes through each proposition one by one and determines whether that proposition is true. When there are millions of possible answers the process can be time-consuming. A neural net making a decision doesn't consider whether these individual propositions are strictly true or false. Instead, each proposition has *weight,* which might be characterized as the strength of the network's "opinion" as to whether it is true or false.

One way to understand this difference is to think of the two problem-solving methods geometrically. Suppose that for a particular problem there are only two propositions to choose from, each either true or false. Each corner of a square represents a possible state of the two propositions. One corner of the square can be labeled 0,0: both propositions are false. The next corner is labeled 1,0: the first proposition is true, the other false. The next corner, 0,1: the second proposition is true, the first false. The last corner is labeled 1,1: both propositions are true. The four possible true/false states of the two propositions can be represented as corners of the square, and the only place those propositions logically make sense is on those corners. (See the Figure on the following page.)

A typical computer would solve the problem by starting at one corner, testing the truth of the propositions, then moving to each of the other corners. The process is quick if there are only four corners to examine, but for a complex problem with millions of "corners" representing possible solutions, the process can take so long that the problem is practically unsolvable.

A neural network starts, not at a particular corner but within the *interior* of the square, then moves outward until it gets to a corner where a logical answer appears. Even though the corners of the square are the only places with logical meanings, the network finds the answers more effectively by moving around in the illogical interior space. "It's amazing," says Hopfield. "The network is better at these kinds of problems because it looks for the answer by moving around in the interior—a place where there are no answers." By starting in the interior of the square the network is, in a sense, considering all the answers at once. As it moves closer and closer to a smaller number of corners, more and more of the incorrect solutions disappear, and the network quickly arrives at the right one.

This kind of problem solving is typical of the tasks our brains must perform in everyday life. "In biology, an awful lot of problem solving has to be done on the first pass," says Hopfield. "For example, if I'm talking and you don't understand a particular word I said, you can't just keep the sound of the word reverberating in your head and try to sort it out. Because I'm continuing to talk. You have to get everything on the fly."

A neural network's brainlike problem-solving abilities have excited many researchers who have long been trying to simulate

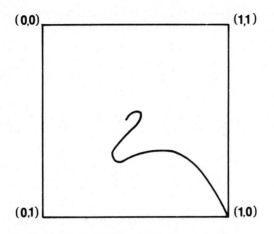

Neural networks solve problems in a manner different from conventional computers. One way to understand this difference is to imagine a hypothetical "answer space" containing all the potential solutions to a particular problem. The computer's task is to find the point in this space representing the correct answer. A problem with four potential answers, for example, could be thought of as a square with an answer at each corner. A typical computer will attempt to solve such a problem by starting at one corner and testing whether the propositions there are true or false, then traveling to each of the other corners until it finds the correct answer. With complex problems, however, the potential answers may number in the thousands or even millions, making this type of step-by-step journey to each "corner" in the imaginary answer space extremely time-consuming. A neural network, on the other hand, can be thought of as starting its search somewhere in the middle of the square, where there are no potential answers. Instead of testing for whether an answer is absolutely true or false, it gropes towards those corners that seem more correct than others, eventually making its way towards the best match. For complex problems this type of search may prove to be faster than the step-by-step approach.

human intelligence in machines. But using a neural network to solve problems involves some tradeoffs. "I'm not a hundred percent sure people would really tolerate machines with too many humanlike qualities," says Hopfield, "because in some sense, having a machine with a degree of creativity means that you are also giving up a degree of control—and I don't know whether people are going to be willing to give up that control."

The Creative Computer

Conventional computers are intentionally designed so that unexpected behaviors don't emerge from the interactions of the various electronic components. Neural networks, on the other hand, get their power from these interactions. But, as a result they also display a few maverick characteristics. For example, besides storing memories deliberately entered, a network will sometimes create its own spurious memories. The memories don't come completely out of the blue; they're usually the result of the network making new correlations between parts of other stored memories.

Suppose you are using a neural net to store the recreation schedule for three people. So you enter:

John–tennis–Wednesday
Phyllis–golf–Wednesday
Dick–golf–Thursday

These memories are distinct, but they also have some parts in common. When a neural net is given these memories, it may also create a new memory—*John–golf–Thursday*—even though that one wasn't put in. *John* has been correlated with *Wednesday,* and *golf* has been linked to *Wednesday* and *Thursday,* so the network creates a new arrangement. In a sense, the memory is perfectly reasonable. It's just that this particular memory isn't real.

Hopfield has found a way to eliminate these spurious memories from a neural net. Since its memories are like valleys in an "energy terrain," you could think of the stronger memories as having deeper valleys than those of the weaker memories. The way you even out those memories, says Hopfield, is by wandering randomly through the network, slightly "pushing up" the bottom of all the valleys you fall into. Through this process, the memories are evened out. "It's an *unlearning* procedure," says Hopfield. "We call it *negative learning,* because you are undoing the valleys a little to even them out. It turns out that this procedure greatly reduces the numbers of spurious memories, because those memories tend to be very weak."

When we dream, our brains may be doing a similar type of "unlearning," says Francis Crick, who works on brain science at the Salk Institute in San Diego. Crick and his colleague Graeme Mitchison suggest that during the day, our brains soak up all kinds of stimulations and associations, which overwhelm our brains' neural networks and create spurious memories. These spurious memories are linked together and become fantasies and dreams. As we sleep, says Crick, we replay those memories and associations; the legitimate memories are strengthened, while the spurious associations are reduced. "We dream in order to reduce fantasies and obsession," says Crick. "Attempting to remember one's dreams should perhaps not be encouraged, because such remembering may help to retain patterns of thought which are better forgotten." But having spurious associations in a neural net may not be all that bad, because putting things together in a new way is the basis of creativity. "If you want to have a new behavior, what you'd call *originality*," says Hopfield, "this is a way to generate it."

Much of the growing interest in neural networks stems from another mindlike trait that the networks only recently have been given, an ability that has opened the way for dramatic improvements in the capabilities of the new models of the mind. Networks like Hopfield's can make only simple associations between input and output, but a recent theoretical breakthrough has given neural networks a new talent—the ability to program themselves to perform sophisticated tasks by learning through experience.

CHAPTER SIX

THE SELF-TAUGHT COMPUTER

Neural Networks Learn to Learn

An idea is a feat of association.
—Robert Frost

It takes a keen eye to know what's a dead end, and what's a false start.
—Robert Craft

Jay McClelland is scrambling around the kitchen trying to get dinner ready for his wife, Heidi, and their two children. He's running late. He's spent too much time talking to a reporter at his office and finally decided that the only way to break free was to invite the reporter to dinner at his home near Carnegie-Mellon University. On the menu tonight are tacos. The meat is thawed and cooked in the microwave oven, and the tortillas are also steamed in there. "There is one advantage to being a connectionist and also a parent," says McClelland, stirring the ground beef as steam rises from the plastic bowl. "You don't constantly worry about whether you're teaching your kids the right rules all the time. You just have to give them good examples. Neural networks—and kids—are perfectly capable of learning the right rules on their own."

The Neural Net Revolutionaries

McClelland is probably one of the best-known psychologists in his field, though *notorious* might be a more accurate description. Along with David Rumelhart, of Stanford University, the University of Toronto's Geoff Hinton, and Johns Hopkins' Terry Sejnowski,

McClelland is one of the main architects of the new neural net revolution. Some cognitive psychologists have welcomed the revolution's new model of the mind; some hate it. But few are indifferent to it, and it's a good bet that most of them know of McClelland's work.

In 1986, McClelland and Rumelhart published a two-volume book entitled *Parallel Distributed Processing: Explorations in the Microstructure of Cognition*. It sold out its first printing of five thousand copies in one week. A sprawling 1200-page tome outlining the basic neural net program and giving concrete information on modeling, *Parallel Distributed Processing* may become the connectionist's bible—except that at the speed that connectionism is catching on in the cognitive community it may soon be out of date. A third volume was released in 1987.

The title page of Volume 1 lists Rumelhart first as editor; in Volume 2, McClelland is first. In both volumes, the authors' names are followed by another credit: "and the PDP Research Group." PDP stands for parallel distributed processing, a more precise term for the dominant model of connectionist architectures: *parallel* because all the neurons act at once; *processing* because the system is not just a simple memory device, but also uses information to go from a particular input to a particular output; and *distributed* because no single neuron is responsible for any particular function, the activity being spread among many neurons.

The fact that these networks are also called *connectionist* models is due to the efforts of one of the first of the new generation of neural networkers, Jerry Feldman, who popularized the term. Feldman, a computer scientist at the University of Rochester, works on a type of neural network that uses "local" representations. That is, individual neurons are used to represent specific entities, such as one neuron representing the color blue or a concept like "large." These representations can be combined to create theoretical structures of how we think, but since it's unlikely that there is a specific neuron in your brain responsible for representing the concept of, say, your grandmother, most connectionists have turned to network models that are distributed—and thus more brainlike—types of representations.

The PDP group meets regularly at the University of San Diego, where McClelland, Rumelhart, and Hinton have all worked, and includes psychologists, computer researchers, philosophers Patricia and Paul Churchland, and a handful of linguists, anthropologists, and sociologists. Francis Crick also contributes, as do researchers from afar like Sejnowski. As Rumelhart and McClelland point out in the Introduction to *Parallel Distributed Processing*, their book "reflects the influences of the group. We hope the result reflects some of the benefits of parallel distributed processing." The PDP group, in other words, processes information in the same manner as the connectionist networks they study.

The Interactive Mind

Like many neural net revolutionaries, McClelland began his career as a mainstream cognitive scientist but grew increasingly frustrated with the old model of the mind. "The dominant view was that the mind was a very sequential, discrete-stage computing device," he says. "I was trying to use this view to understand how *context* influences *perception,* but I just couldn't make it work from this stepwise point of view. There were these contradictions I kept coming up against that made it very difficult."

For example, suppose that you are looking at something like the Figure below.

Even though your brain can't completely identify the letters, it makes a "best guess" to recognize the word.

You can't make out the first letter of the word; it's either *R* or *P*. The second could be *E* or *F*. And the last letter could be a *D* or *B*. The standard psychological theory, says McClelland, was that when you recognize words, you first recognize the individual letters in the word, then put those letters together to recognize the word as a whole. "But if you do that here," he says, "you get stuck. You *can't* recognize the letters first—they're ambiguous. Yet when you look at it, you can see that the word is *RED*. How can you recognize the word before you can recognize the letters? That was the dilemma that I was faced with."

The solution, says McClelland, is that instead of having an initial step where you recognize the letters, you instead have a mechanism that indicates its *degree of confidence* in the different possibilities the letter might be. The mechanism might say that there's evidence that the first letter is *R,* and there's also evidence that it's *P,* but there isn't much evidence of anything else. It then performs the same process on the other letters, and all the information is sent to another level of processing. This level has various *word* detectors, and the one that looks for the letters R-E-D becomes activated because it sees the possibility for a perfect fit.

None of the other word detectors is excited because there is no other fit, and so the device concludes that the word is RED. "It seems like an extremely natural way of accounting for this contextual effect," says McClelland. "One layer of processing provides a range of options. The next level of processing looks at this output, and makes a decision among the possibilities that it's willing to consider." Working from those ideas, McClelland developed what he calls the "cascade" model of word recognition and joined the ranks of connectionists.

Neural Nets' Troubled Past

It took a good deal of courage—some might call it foolishness— for any scientist to go into connectionism during the 1970s. A connectionist movement had already taken place in the late 1950s and 1960s and was considered a failure by most scientists. This neural network movement was pioneered by researchers Bernard Widrow and M. E. Hoff, who experimented with a machine called the *adaline* that used networks of adaptive neuronlike units to recognize patterns.

Another neural network pioneer was a psychologist from Cornell University, Frank Rosenblatt. In 1959, Rosenblatt built a simple neural network machine that he called a *perceptron*. The perceptron had a grid of four hundred photocells, arranged a little like an eye's retina, that were randomly connected to 512 neuronlike units. When a pattern was displayed to the sensory units—for example, an *A* or a *B*—the sensors would activate one set of neurons, which sent a signal to another bank of neurons that indicated whether that pattern fell into the *A* or *B* category. By adjusting the layer of connections between the neurons, Rosenblatt was able to get his network to recognize all the letters of the alphabet.

Like the research into digital computers that was taking place at the same time, explorations of the perceptron and other neural net machines produced quick and tantalizing results. As a consequence, supporters of neural networks fell prey to a wide-eyed optimism similar to that of their counterparts working with digital computers. "For the first time, we have a machine which is capable of having original ideas," wrote Rosenblatt. "As a concept, it would seem that the perceptron has established, beyond doubt, the feasibility and principle of non-human systems which may embody human cognitive functions at a level far beyond that which can be achieved through present-day automatons. The future of information processing devices which operate on statistical, rather than logical, principles seems to be clearly indicated."

These two competing approaches to creating an artificial mind—neural networks and digital computers—soon became the focus of an often acrimonious debate among researchers in the artificial intelligence community. The debate centered on a simple question: Which is more important—using symbols and logic to try to simulate what the mind can do or using neuronlike hardware to try to simulate what the brain does?

In the center of the debate was MIT's Marvin Minsky, one of the most influential computer researchers of the last three decades. Minsky was one of the first people to experiment with a type of machine learning that had much of the flavor of neural networks. In 1951, he built a machine with forty adaptive units that were linked together and learned by reinforcement. Minsky's Ph.D. dissertation was the first thesis dealing with neural networks to be written at Princeton University. (The next graduate thesis on the subject at Princeton came some twenty-five years later, written by Terry Sejnowski.) But Minsky, like many other researchers at the time, was lured to the great potential of digital computers. In the early 1960s, digital computers were rapidly becoming larger, faster, and cheaper, and the rule-based approach to problem solving quickly yielded results.

In 1969, Minsky and a coworker at MIT, Seymour Papert, published a book, *Perceptrons,* which gave a mathematical analysis of just what simple neural net machines could and couldn't do. The book was also quite critical of the neural net movement. "Perceptrons have been widely publicized as 'pattern recognition' or 'learning' machines," Minsky and Papert wrote, "and as such have been discussed in a large number of books, journal articles, and voluminous 'reports.' Most of this writing . . . is without scientific value."

Minsky and Papert demonstrated that perceptrons were fundamentally incapable of doing simple tasks such as determining whether the number of spots on a grid was odd or even and whether a particular figure was connected or not. The book's mathematics was indisputable, and its tone gave the message that perceptrons were a very dead end. Connectionism quickly went into scientific hibernation.

In the current revival of connectionism Minsky and Papert are often portrayed as the evil huntsmen who slew Snow White. But the researchers' main goal, says Papert, simply had been to give an objective scientific assessment of the potential of perceptrons. There were other motivations besides intellectual curiosity, however: Government agencies were beginning a flood of funding for research in artificial intelligence, and researchers were scrambling for a piece of the pie. "By 1969, the date of the publication of *Perceptrons,* AI was not operating in an ivory-tower vacuum," notes Papert. "Money was at stake. Part of our drive came, as we quite plainly acknowledged in our book, from the fact that funding and research energy were being dissipated on what still appear to me to be misleading attempts to use connectionist methods in practical applications."

A Simple Neural Net

Perceptrons are an example of the simplest type of neural nets. They have one layer of modifiable connections between their input and output neurons, and knowledge is represented as patterns of connections among all the neurons. Information is shuttled through the network by the actions of neurons signaling to one another.

For example, suppose you wanted to design a machine to tell you whether to serve red or white wine with your grilled T-bone steak or poached haddock. A program for a computer would be fairly simple: Step 1 might be "If it's steak, then go to step 2; otherwise, go to step 3." Step 2 would be simply the order "Serve red wine." Step 3 would be "Serve white wine."

You could also design a neural net to do the task using four input neurons and four output neurons. (See the Figure below.) Where each of the input neurons crosses each of the output neurons, a connection can be made. When the connection is switched

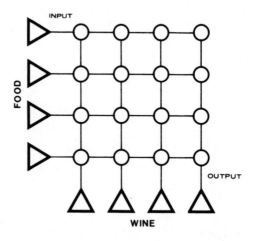

Training a neural network: In this simple network, the vertically arranged triangles represent input neurons and the horizontal triangles are output neurons. A filled-in circle on the grid represents a connection between an input neuron and an output neuron. The network is trained by giving it an input pattern representing a particular kind of meal, for example, then setting the desired output pattern for a particular beverage, and adjusting the connections between neurons. After training, the network can make the "food-and-beverage" associations by itself.

on, the two neurons are communicating with each other (a connection is shown in the diagram as a filled-in circle). When the connection is off, no information flows through the connection (a non-connection is shown as an empty circle).

For this particular neural net you represent steak as the string of digits 1010; fish is 0101. That means, for steak, the first and third input neuron is switched on. For fish, the second and fourth neuron is switched on. Active neurons for each dish can be assigned arbitrarily, but you could also think of each neuron as a "feature detector." That is, the first neuron could be sensitive, say, to the presence of dark grillmarks. The second neuron could look for light-colored meat. The third neuron could be sensitive to reddish meat. And the fourth could be designed to switch on in the presence of tiny bones. When grillmarks and red meat appear, the first and third neurons switch on. Since there are no tiny bones or white meat present, the other two neurons remain inactive.

Now, how do you make the network produce the right kind of wine for the right meal? Unlike computers, neural nets aren't "programmed." Instead, the connections between the neurons are adjusted so that the input of a particular pattern—in this case, the pattern representing steak or fish—produces an output representing a particular wine. For this simple example, red wine is represented by 1100, white wine by 0011.

First, you enter the pattern for steak into the input neuron by switching on neurons 1 and 3 and put in the output pattern you want the net to produce—red wine—by switching on output neurons 1 and 2. (See the Figure below.)

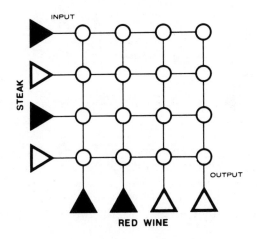

The next step is to adjust the connections in between. Since the input neuron 1 and the output neuron 1 are both on, you switch on the connection between those two neurons. You do the same thing for each juncture where you find a pair of input and output neurons on at the same time. (See the Figure below.)

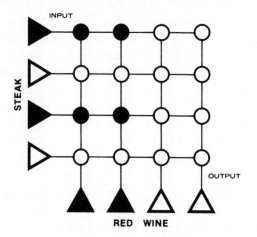

Next, do the same thing for the input pattern of fish and the output pattern of white wine. When the network has been trained, it should look like the Figure below, where each filled circle means a connection between an input and output neuron.

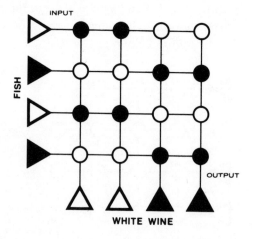

You can see from the network that once it's set up, whenever it's given the "steak" pattern 1010, the first input neuron activates its "row," as does the third input neuron. Since the training session produced connections only between the first and third input row and the first and second output column, only the first and second output neurons fire, producing the pattern 1100 for red wine. Likewise, when the input pattern for fish is presented, the connections are such that only the output pattern indicating white wine will fire.

The network will also do a simple version of "best guess." Given a partial pattern for steak, 10xx, the network will still produce the 1100 output pattern for red wine.

What about pizza? If you want the network to tell you that for pizza the beverage of choice is beer, you install that pattern as well. Using 1100 as an input pattern for pizza, and the output pattern 1001 for beer, you add a few more connections to the neural net. (See the Figure below.)

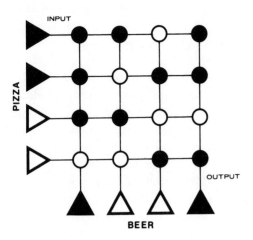

And of course, for those who prefer a nice cool glass of milk with their peanut butter and jelly sandwich, that association can also go in as well. Using 0011 as input for the sandwich and 0110 for the output pattern for milk, you add even more connections. (See the Figure on the following page.)

Putting in these extra patterns creates some problems, however. You can see that if the pizza input pattern 1100 is presented, some current will flow into *all* the output neurons, producing the meaningless output pattern 1111.

To make the network operate correctly you have to make its output neurons a little more like their biological counterparts; you give them a *threshold*. That is, you stipulate that they will not

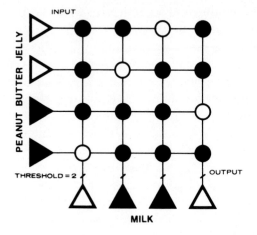

switch on unless they get not one but *two* signals from the input neurons. Thus, when 1100 is the input, current flows to all the output neurons, but only two output neurons—the first and fourth—get enough of a signal to turn on. This type of neural net, explored by David Willshaw, of the University of Edinburgh, during the early 1980s, is one of the most basic. Like Hopfield's design, it was one of the first to be explored by neural net chipmakers.

Obviously, the food-and-drink neural net is quite limited; it can hold a few patterns and make some simple associations. But more sophisticated neural nets have been created by embellishing that simple design. There are nets made with many more neurons, nets that use a variety of different strategies for connecting the neurons, and nets with neurons that signal in more sophisticated ways such as using plus or minus one or a sliding scale of numbers such as 1, 2, and 3 or incremental values such as 0.25. A network's neurons can also have higher or lower thresholds. These sophisticated nets can have an associative memory—that is, they can take a partial input and fill out the rest of the output pattern.

The early researchers soon realized, however, that as a network became larger and more complex it became harder and harder to determine the proper values for the connections between the neurons. So researchers began looking for the best way to make neural nets adjust the connections between neurons by themselves. In other words, the nets would have to *learn*.

The simplest types of learning procedure for a neural net have been inspired by Hebb, who proposed that biological minds learned because the communication links between two active neurons became stronger. Hebb's initial ideas were the backbone for many learning schemes devised by the early neural net researchers as well as the new connectionists. These learning procedures have different mathematical twists, but at their essence is Hebbian simplicity: A neural network is trained by giving it a pattern of activation among its input neurons and, at the same time, setting the desired pattern among the output neurons. Then the connections between neurons are adjusted. If a particular pair of neurons contributes to making the correct association between the input and output patterns—that is, if they are both active—then the connection between the two neurons becomes stronger. With this simple type of learning, a neural net can be trained to recognize patterns and make generalizations.

As Minsky and Papert so convincingly demonstrated, however, simple neural nets are fundamentally incapable of performing some tasks. Computers, on the other hand, are universal machines; in theory, they can compute *anything*. Faced with the choice between studying the two very different designs for thinking machines, one that appeared to have serious limitations and one whose potential seemed endless, most scientists abandoned work on neural nets.

But perhaps because the research community was caught up in the heat of Minsky and Papert's critical rhetoric, scientists overlooked something in their analysis of perceptrons. It took nearly fifteen years for scientists to discover what that was, and to realize that it was dark at the end of the neural net tunnel only because the path took a slight bend.

The Keepers of the Flame

During those years, there were still a handful of scientists who carried on with neural net research. Despite what theorists said about the limitations of perceptrons, all the evidence from neurobiology suggested that the brain was *some* sort of neural net. Ironically, one more bit of evidence that a brain is more like a perceptron than a computer came from the cover of Minsky and Papert's book. On that cover are two spirals (see the Figure on the following page). One of the spirals is made with a single line; the other is actually two lines nested within each other. Which is which?

Most people can't immediately see which is connected just by looking at the two spirals. To determine which is which, you must probably perform some kind of stepwise, computerlike operation such as tracing the lines with your finger. So while Minsky and Papert rejected perceptrons as models of the mind because of their inability to do simple tasks like recognizing whether a particular pattern is connected, our brains also have trouble determining which of the two spirals is connected at first glance—just like perceptrons. Only after a lengthier period of cognitive work are we able to tell the difference between the two. A perceptron obviously has limitations because it can't recognize connectedness at all. But in some ways, its behavior is very much like our brains.

One scientist who continued to work on neural networks during those dark years was Stephen Grossberg, now the director

One of these spirals, which adorn the cover of the book *Perceptrons,* is made with a single continuous line. The other is made with two lines nested within each other. Which is which? In *Perceptrons,* Marvin Minsky and Seymour Papert pointed out that one limitation of neural networks is their inability to determine whether a figure is connected. As demonstrated by these spirals, our brain sometimes has difficulty doing that task as well.

of the Center for Adaptive Systems at Boston University. Grossberg is working with Gail Carpenter of Northeastern University on a neural net model called *adaptive resonance theory,* or ART. Their most recent network, ART 2, combines "top-down" and "bottom-up" processing to categorize identical drawings of trucks, for example, even if they are different sizes and orientations. Other researchers who continued to work with neural networks during the 1970s were Teuvo Kohonen of Helsinki University and Brown University's Jim Anderson.

Anderson looks like the kind of person who might make a career out of doing things he was told not to do. A large, soft-spoken man with a long beard, he remains on the edge of things, even among the new connectionists. Anderson helped to nurture the neural network field and brought some seeds of the new connectionist movement to the University of California at San Diego toward the end of the 1970s. There, he and Geoff Hinton organized what could be thought of as the first "neo-connectionist" meeting in the summer of 1979. Soon afterwards, Anderson and Hinton edited a collection of papers from the meeting that dealt with neural networks and associative memories.

Today, many of the people who attended that first meeting have added new architectures and learning methods to neural network models. Other researchers have joined the ranks, too. Across the campus at Brown University works Leon Cooper, a physicist who won a Nobel prize for his work in superconductivity and now studies the brain. Cooper is the founder of Nestor, Inc., one of the first companies to offer commercial applications of machines that use neural net designs. Anderson, however, has stayed closer to the lines of his original work, refining his models and designing new experiments to test them, largely leaving the task of breaking new ground to those he helped pull into the field.

Anderson's perspective comes from his training in neurophysiology at MIT, where he first tried to understand how neurons work as a group. "The only serious drawback with that approach," he says, "is that it tends to get you unemployed. Back then, studying the brain as a *system* wasn't a real high status way of proceeding. Instead, a neuroscientist is supposed to get an animal, drill a hole in its head, stick in an electrode, and see what connects to what."

For his graduate thesis, Anderson followed the traditional neuroscience lines. He worked on the snail *aplysia,* which has a simple, well-defined nervous system. He then went to the University of California, Los Angeles (UCLA), for more research, but soon after he arrived, the researcher left the university, and Anderson suddenly found himself in a position to do whatever he wanted. He wrangled an office at the brain research institute and began reading whatever he could about neuroscience. "It was a great experience," he says. "I viewed myself as the audience—you know, somebody *writes* all these papers, and somebody has got to *read* them and figure out what it all means. I figured that was my function."

Anderson began to forge his own ideas about connectionism. It struck him that in the brain, there are no special compartments for storing different memories. "Human memory is not a filing cabinet," he says. "Memories are spread out and mixed together when they are stored." At first, Anderson wondered why the brain would use this method of memory storage; it seemed that memories would overlap and interfere with each other. But then he realized that, in fact, there might be advantages to having such overlapping memories. In a filing-cabinet type of arrangement, no one memory is more important or accessible than another. But in an overlapping storage system, those memories that are similar to each other are stored together and are reinforced. Therefore, as you accumulate memories, things that are familiar should naturally become stronger memories. "I found it amazing," he says. "You could create a neuroscientific model with neurons and synapses, and the model could make predictions about what should be happening *psychologically.*"

Anderson began studying how different memories are stored and retrieved with his own models of neural networks. He didn't know that he was a connectionist until several years later. "I was at a cognitive science meeting in 1984, and everybody was talking about connectionist modeling," he says. "And I thought to myself, 'Hey, that must be what I'm doing.' "

The general idea of connectionism goes back as far as the eminent nineteenth-century psychologist William James. James, in his *Briefer Psychology* published in 1890, wrote:

> The more other facts a fact is associated with in the
> mind, the better possession of it our memory retains.
> Each of its associates becomes a hook to which it hangs,
> a means to fish it up by when sunk beneath the surface.
> Together, they form a network of attachments by which
> it is woven into the entire network.

It wouldn't be accurate to characterize James as one of the first connectionists, but it is clear that his notion of memories being reinforced by associations and woven into a network has the flavor of connectionism. "It's not as though there are any big deep-dark secrets that have to be discovered about neural nets," says Anderson. "That's why a lot of these basic ideas have been reinvented by many people—not exactly 'reinvented,' but they tend to appear in different contexts."

Through Anderson, the old connectionist, the new connectionist McClelland began learning about neural networks. McClelland met Anderson while both were attending a conference in 1975. Anderson showed McClelland some of the neural network models he was working with. "As I listened to him talk about some of the equations he was using in his models, it occurred to me that they were just dead naturals for capturing some of my ideas of continuous output," says McClelland.

McClelland began trying to develop his own neural network models at the University of San Diego, where he was an assistant professor. McClelland had little experience working with computers—that was before every scientist had one resting on his or her desktop—and the university's computer was an early model. It was programmed by feeding a stack of punched cards into it, and took two days to produce results. "I had a horrible time with that stuff," says McClelland. "I just couldn't get the program debugged. So instead of continuing to do simulation models, I decided to develop a mathematical formulation of my theory. I can learn mathematics if I have to, and I can use it as a tool. But I don't think like a mathematician. I simply learned the mathematics I needed to do the work."

The Building of the Neural Net Revolution

McClelland looked for help with mathematics from another cognitive science professor, David Rumelhart. McClelland sat in on a course Rumelhart taught in mathematical psychology. These were freewheeling sessions where Rumelhart would get up and say, "Okay, what are we going to develop a mathematical model of today?" Graduate students would raise their hands and describe the behavior of their pigeons or mice in their laboratory experiments. Rumelhart would scratch his head for a few minutes, then begin writing equations, erasing them, and trying other approaches. Sometimes he'd stop and say, "Well, I don't know how to prove this one," and go back to his office and get a book. Then he'd come back and start writing new equations. "Rumelhart never gave up—he'd just work it through," says McClelland. "By the end of the semester I said to myself, 'By God, if he can do it, then I can do it, too.' That's where I got the courage to do mathematics."

At the time, back in the late 1970s, Rumelhart was not thinking about connectionism. A tenured professor, Rumelhart had made a name for himself in another area of cognitive science. Rumelhart had been studying how people can fill in the details of a story without being explicitly told everything that is going on. Consider these two sentences: Mary heard the ice-cream truck coming down the street. She remembered the money she got for her birthday and ran into the house.

Somehow, we know that Mary is a little girl and that she is going to use her birthday money to buy some ice cream. We are able to fill in all that missing information ourselves, drawing on our own resources to fill out the story from the little information we are given. Change the words *ice-cream truck* to *armored car,* and *money* to *gun,* and you would conjure up an entirely different scenario.

The words we generally use when we talk to each other are actually so threadbare of information that most of the work in a typical conversation is done by the *listener,* not the speaker. The listener is continually taking cues from the speaker and applying them to a kind of prepackaged memory that helps determine what is being talked about. A listener has to know the speaker's expectations. When someone asks you, "Do you know what time it is?" you don't answer, "Why yes, I do."

Some scientists, such as Minsky, call these memory packages *frames*. Roger C. Schank, of Yale University, calls them *scripts*. Rumelhart calls them *schema*. Trained in mathematics, Rumelhart soon found that he needed a more powerful language to develop his schema models, and he was drawn to the then emerging field of artificial intelligence. AI researchers had sophisticated formulations for talking about the nature of memory, the mind's internal structure, and the representation of knowledge.

Rumelhart became particularly interested in the work of artificial intelligence researcher Ross Quillian, who was working on the problem of how knowledge might be organized in the mind. Quillian proposed that if two words have similar meanings, such as *bird* and *canary,* the words should be related in memory. Quillian developed the idea of a *semantic network,* a web of knowledge arranged according to how objects are related to each other.

The idea of a network of associations led Rumelhart to become interested in what he calls *cooperative computation.* To most researchers in artificial intelligence, there is a distinction between "bottom-up" and "top-down" systems. A bottom-up system starts from individual parts to create a whole; the tiniest pieces of a visual scene, for example, are put together into bigger and bigger pieces to construct an entire image. A top-down visual system works in the opposite direction, starting with an overall idea of what it expects to see, then breaking that idea into smaller pieces that it tries to fit to the data in the real world.

Since there is evidence that our brains use both processes, Rumelhart began to explore the idea of a system that would combine top-down and bottom-up, working simultaneously—an interactive, cooperative method of processing information similar to McClelland's "cascade" model of word perception. In Rumelhart's system, data would come in, suggesting possibilities; then another level of processing would check whether data from the world really conform to any of those possibilities. The system would use different sources of knowledge, all working simultaneously to create an interpretation. Many researchers objected to Rumelhart's interactive processing ideas, however, claiming that the brain worked too slowly to implement such a model. That spurred Rumelhart toward making his models more biologically plausible, to show that they could conceivably be carried out by brains.

In the late 1970s, Rumelhart began working with McClelland,

on an interactive model of reading. During their four-year collaboration, Rumelhart and McClelland became close friends; Rumelhart's low-key style meshed with McClelland's buoyancy, and both enjoyed thrashing out ideas in conversation. Even now, though the two researchers are at either end of the continent, they work together on problems whenever they meet. At meetings of cognitive scientists, they can be seen huddling over a cup of coffee in the cafeteria, frantically scribbling on paper napkins. Many of the ideas that they put into *Parallel Distributed Processing* were originally sorted out at such sessions or standing at a blackboard, working through months of conversation until they had fully explored their theories.

Another influence on Rumelhart's conversion to connectionism was his collaboration with Don Norman, the director of the Cognitive Science Department at UCSD. Norman and Rumelhart were interested in motor control and began by studying how people type. Some people can type very quickly; good typists work with only about a fifty-thousandth of a second between strokes. How can they go that fast?

At first, when the two researchers looked into the phenomenon, other scientists were skeptical. Most researchers in AI didn't consider cognitive science to include the brain's control of body movements. Neuroscientists, however, were intrigued by the idea that people can type so quickly, because fifty thousandths of a second is barely enough time for a signal to go from the brain to the finger and back again. To find out how people were able to move their fingers that quickly despite the slowness of the brain's signaling powers, Norman and Rumelhart took high-speed movies of people typing. Analyzing the films, they found that people overcame this limitation in speed by essentially typing in *parallel*. When someone types the word *vacuum,* for example, his left index finger goes down to push the *v* at the same time that his right index finger is moving up to the *u*. That's four characters in advance. In other words, when you type a word, you basically type the whole word at once; all the fingers are reaching out for all their letters at the same time.

Of course, if one finger has to type two letters in a row, it can't move to both of them at once. But Rumelhart and Norman found that when people type a *v* followed by an *e*—which are typed with fingers of the same hand, but on different rows of keys—they distort the hand so that they reach for both keys at once. Rumelhart and Norman's model of typing, which mirrored this parallel method of processing, moved Rumelhart closer to connectionism.

In 1978, Geoff Hinton came to UCSD from England. Hinton is a brilliant scientist with a quick wit and a penchant for wearing outrageously patterned shirts. His interest in connectionism goes back to when he was a teenager. "I was thinking about neural nets when I was sixteen," he says. "I had just found out about holograms, and it seemed to me to be an obvious metaphor for the mind. I was trying to find a way of using a vacuum molder—a machine that uses heat to mold plastic—to make neurons out of water and plastic. I figured that if I could get my hands on a vacuum molding machine, I could simulate neurons in a network using funny-shaped pipes and water running through them. I didn't know about computers then."

Hinton had worked on traditional artificial intelligence and vision at Cambridge University, and later at the University of Edinburgh. But he was committed to connectionism, and when he saw a notice of a postdoctoral position opening up in cognitive science at UCSD, he decided to give it a try. At UCSD, Hinton began working with Jim Anderson on creating associative memories in neural networks. Rumelhart knew of Hinton's work on associative memory, but he didn't see a relationship between that and his interactive-reading and parallel-typing models. When Anderson and Hinton organized the first neo-connectionist meeting in 1979, however, things began to gel. The meeting included UCSD's Rumelhart, McClelland, Hinton, and Anderson, as well as Jerry Feldman and Terry Sejnowski. From this meeting, the field of neural networks was reborn.

After that first meeting, however, things progressed slowly. Hinton returned to England, and Rumelhart went to Stanford for a year as a visiting professor to write a book on his ideas about schema. But he couldn't keep his mind off connectionism. Slowly, Rumelhart began to see how the various neural net models and interactive processing ideas were related. He returned to UCSD in 1981, convinced that he had to spend some time sorting out his connectionist ideas.

When Geoff Hinton came back to UCSD from England, he joined McClelland and Rumelhart in a six-month effort to organize their neural network ideas. In the end, they vowed, they would write a book. They met twice a week, eight hours a day, and soon other researchers like Francis Crick began attending the meetings. But after months of intensive work they had barely scratched the surface.

It became clear to Rumelhart that he would need at least five years to fully develop his ideas about neural networks. That posed a problem. Rumelhart had made his scientific reputation on his schema ideas and had intended to write a book on the subject. That book had yet to be written, and now this new field of neural networks was opening before him, a field that had fizzled once before. "I remember very well having a discussion with my wife," says Rumelhart. "Should I write that book on schema, or should I dig some new ground? Should I donate my next five years to something that might be a losing cause? My wife and I decided that, yeah, go for it. So I put away the book and went off to explore these new ideas."

The Power of Multiple Layers

One of the new ideas Rumelhart and his colleagues explored has now become the backbone of the connectionist revolution, because it gave neural networks a sophisticated learning ability. In *Perceptrons,* Minsky and Papert focused their analysis on neural nets made with a single layer of connections between the input and output neurons. Toward the end of the book, the authors briefly examined perceptrons with multiple layers of connections. They saw no reason to suppose that multilayered networks would perform any better than single-layered nets and concluded that it would be an important research project to "elucidate (or reject) our intuitive judgment that the extension is sterile."

A few researchers had been experimenting with multilayered nets at the time, but no one had been able to find an effective way to set up the connections between the neurons in the various layers. On simple networks, setting up the connections is relatively easy because the settings for the input neurons and the output neurons are already known; they match the input or output of the task the network is supposed to do. But on multilayered nets, the neurons in the middle layers are not part of the input or output, so it's difficult to connect them. Because of these difficulties, the learning rules for multilayered nets that existed back in the 1960s were slow and ineffective. And when Minsky and Papert's book appeared, the idea of multilayered nets died along with the simpler perceptrons.

The new connectionists have rediscovered multilayered nets. A multilayered net can do what Minsky and Papert showed a single-layer perceptron can't do, and it's capable of much more. One of the limitations of single-layer neural nets, for example, is their inability to deal with the "exclusive-OR" problem. An exclusive-OR problem arises when someone is given a choice between two alternatives and there is an equal preference for one or the other, but not both of them.

For instance, suppose your romantic friend Patrick has been dating Susan and at the same time dating Pam. He likes both women and wants to continue both relationships. You are hosting a party, and you know Susan and Pam. Should you invite neither of the women, one, or both? For Patrick, the dilemma is a case of exclusive-OR; if *neither* of the women was invited to the party, he wouldn't have much fun; if *either* Susan or Pam was invited, he would be happy; but if *both* women were invited, he wouldn't have any fun.

Minsky and Papert showed that it is mathematically impossible for a single-layer neural net to learn to reflect this situation. That is, no single-layer net can be trained to produce a 0 in response to neither of the women at the party, 1 in response to either Susan or Pam, but 0 when it gets an input indicating the presence of both. The reason relates to the type of problem that the exclusive-OR poses: At one level of input (that is, inviting either woman), the expected output is straightforward—Patrick will have fun. But at a critical point (in this simple case, when the invitations go to *both* women), the output changes dramatically—from having fun to being very uncomfortable. This kind of problem is too complex for a neural net with a single layer of connections. (Luckily, the problem is easy to solve in real life: Invite both women, and don't invite Patrick.)

Of course, many problems in cognitive science involve more-complicated decision making than the exclusive-OR problem, and so it would appear that being unable to make such simple distinctions is a serious drawback for neural networks. In fact, a neural net *can* make such distinctions, but only if it has more than one layer of connections.

To construct a neural net for Patrick's preferences, you start with an input for Susan and one for Pam, and make your final output whether Patrick will be happy or sad at the party. (See the Figure below.)

Between the input and output you put a layer of so-called hidden units, neurons that are directly involved neither with input or output, but are involved with mitigating the signal. (See the Figure on the following page.)

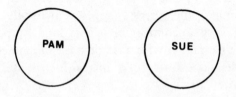

A simple neural network can't model Patrick's dilemma: If either Pam or Sue comes to the party, Patrick is happy. But if both women come—or neither of them—he is not.

Using hidden units, the exclusive-OR problem is easily solved. When either Susan or Pam is the input, the output is 1. If both Susan and Pam—or neither of them—are the input, the output is 0.

Making this type of neural net is trivial, but the effect of adding hidden units to a neural net is not. With hidden units, the limitations shown by Minsky and Papert in *Perceptions* almost disappear. In fact, says McClelland, if the input and output of a problem can be expressed in terms of on and off neurons, neural nets with hidden units have no limitations; theoretically, they can compute *any* type of problem. Like the Turing hypothesis, how-

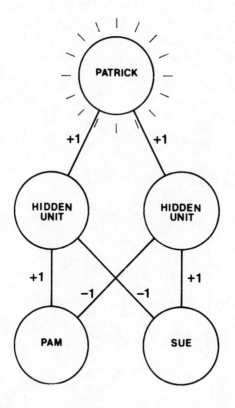

Adding hidden units enables the network to reflect Patrick's preferences. If one or the other input neuron fires, the output neuron fires as well. If both input neurons fire, however, the output neuron does not.

ever, this is a theoretical, not a practical, limit. It remains to be seen whether neural nets can actually live up to their billing.

Hidden units are important because they enable a neural net to form *representations* of the outside world, representations it can use to make complex decisions. In a simple example of this, suggested by William Jones and Josiah Hoskins of the Microelectronics and Computer Technology Corporation, imagine a neural net that is being designed to help Little Red Riding Hood on a walk through the woods.

On her journey, Little Red Riding Hood may encounter three different creatures, and she has several different responses she can make. On meeting a creature with big eyes, wrinkled skin, and a kindly disposition (her Granny), she must learn to approach, give a kiss on the cheek, and offer food. On meeting someone who has big ears, is handsome, and has a kindly disposition (the Woodcutter), she must learn to approach, offer food, and ask for help. And upon meeting someone with big eyes, big ears, and big teeth (the

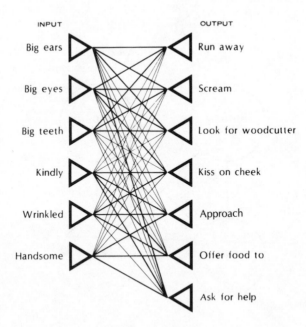

A simple neural network that models the responses Little Red Riding Hood makes as she encounters the Wolf, Grandma, and the Woodsman.

Big Bad Wolf), she must learn to scream, run away, and look for the Woodcutter.

Little Red Riding Hood's decisions are simple enough to be put on a one-layer network, employing the simple learning rules used with early perceptrons. Such a network would look like the one in the Figure on the preceding page.

But you could also design a Little Red Riding Hood neural net that uses three hidden units between the input and output. When the network is set up with the correct connections, it accomplishes the same task. (See the Figure below.)

A Little Red Riding Hood neural net with hidden units uses fewer connections than a single-layer net. More important, each hidden unit represents something in the outside world. The first hidden unit is most active when the input is a creature with big eyes, ears, and teeth. The second hidden unit is most active for someone with big eyes, a kindly disposition, and wrinkled skin. And the third unit is most active for handsome, kindly beings with

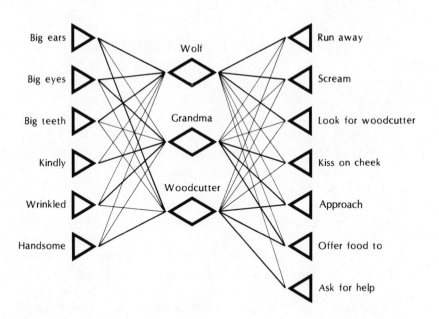

When the "Little Red Riding Hood" neural network is constructed with three hidden units, these middle-layer neurons come to represent the Wolf, Grandma, and the Woodsman.

big ears. The hidden units, without expressly being so designed, become *representations* of the Wolf, Granny, and the Woodcutter.

The ability to form representations gives neural nets with hidden units tremendous power. Hidden units not only allow more-complex types of computation but also enable the networks to represent high-level concepts that are important for complex tasks. "Hidden units bridge the gap between input and output," says McClelland. "They represent what *has* to be represented to solve a problem. They represent *meanings*."

The problem, of course, is determining which are the important concepts and wiring a network to use them. In the Little Red Riding Hood example, the relevant concepts of Granny, Big Bad Wolf, and Woodcutter—and the ways of wiring the hidden units—are fairly obvious. But for a more sophisticated task such as reading a text aloud, the representations needed to solve the problem aren't immediately apparent. Nor is it obvious how to wire correctly hundreds or even thousands of connections in a large multilayered net.

Making Hidden Connections

This was the problem faced by the early neural net researchers back in the 1960s. One of the first efforts to solve it came in 1984 from Terry Sejnowski and Geoff Hinton, who got their inspiration from John Hopfield's theoretical research. Sejnowski and Hinton started from Hopfield's suggestion that when a neural network makes a decision, it can be thought of as trying to lower its energy. But it might get stuck in a valley that isn't the *lowest* in the net, so Sejnowski and Hinton tried to devise a way to move the network toward a deeper valley. "Imagine that you have a model of a landscape in a big box," says Sejnowski. "You want to find the lowest point on that terrain. If you drop a marble in the box, it will roll around for a while and come to a stop. But it may not come to rest at the very lowest point on the terrain. So you shake the box a little and the marble rolls around again. After enough shaking, you usually find it."

Sejnowski and Hinton thought that if they did the mathematical equivalent of "shaking" the neural network, they could make it relax to its lowest energy level. Using this process, their network found the best way of linking its hidden units to its input and output neurons. Their neural network is called the Boltzmann machine after the nineteenth-century Austrian physicist Ludwig Boltzmann who discovered the statistical relationship between the pressure and temperature of a gas and the collisions between its individual molecules.

Hinton and Sejnowski's Boltzmann machine was the first neural network to learn effectively how to set its hidden units on its own. One of the problems Sejnowski and Hinton gave their machine was to tell whether a ten-by-ten array of black and white squares was symmetrical around a horizontal, vertical, or diagonal axis, a problem that Minsky and Papert showed was impossible for single-layer perceptrons to solve.

In their network Sejnowski and Hinton used twelve hidden units that were connected to one hundred input neurons. Each input neuron was responsible for looking at one of the squares in the ten-by-ten array, and three output units were responsible for indicating whether the pattern was symmetrical around a vertical, horizontal, or diagonal axis. Trained on ten thousand patterns, the network eventually learned to categorize any of the 2^{15} patterns possible in that array with about 90 percent success.

At the beginning of training, the strengths of the connections between the hidden units and all the other neurons were set at random. But during training, the hidden units become specialists. Just as the hidden units in the Little Red Riding Hood neural net became representations of Granny, Woodsman, and Big Bad Wolf, the hidden units in the Boltzmann machine came to represent overall features in *their* world. In this case, the units represented the different kinds of symmetry in the patterns the network was trained on.

The network used only as many hidden units as it needed for a symmetry task. In a neural net version of the old saying "Use it or lose it," if the machine doesn't need a particular hidden unit to solve a problem, that neuron simply "withers away." Another interesting feature of the Boltzmann machine, and multilayered neural nets in general, is that there appear to be many ways to set up the hidden units to solve a problem. Sejnowski and Hinton trained their Boltzmann many times, and each time the hidden units arranged themselves slightly differently.

Sejnowski and Hinton used the Boltzmann machine for a variety of problems, teaching it, for example, to distinguish between a *T* and a *C* arranged in any orientation on a grid. But in the long run, it turned out that the Boltzmann machine worked too slowly. Like many neural nets, the Boltzmann machine is simulated on a digital computer, and its operations proved difficult to mimic at high speed. (A real Boltzmann machine, however, may be a lot faster. Josh Alspector and Robert Allen, both at Bell Communications Research, are trying to build a Boltzmann machine on a silicon chip.)

The Boltzmann machine is one of several kinds of learning schemes for multilayered nets in use today. The learning schemes can be roughly divided into three types. In one scheme, called *unsupervised learning,* the hidden units find a way to organize themselves without help from the outside. This method might make a set of neurons compete among themselves for the "right" to respond to a particular input. Eventually, the neuron or group of neurons best suited for particular tasks wins that role.

Another type of learning rule works on *reinforcement* from the outside. Basically, it's a variation of an Easter-egg hunt, where someone is guided by another person telling him that he is getting "hotter" or "colder." In a neural network being trained with reinforcement learning, the connections among the neurons in the hidden layer are randomly laid out, then reshuffled as the network is told how close it is to solving the problem. As the network gets closer and closer to the proper solution, the connections get closer and closer to a workable arrangement. Using this type of learning rule, Andy Barto, of the University of Massachusetts, and Richard Sutton, at GTE Labs in Waltham, Massachusetts, created a neural network simulation of a bug trying to find its way to a tree. Barto used a "scent" from the tree as reinforcement to guide the bug; as the bug got closer to the tree, the scent became stronger. The researchers also trained a neural network to operate the back-and-forth rolling motions of a small cart, so that it could balance a pole upright.

Both unsupervised and reinforcement learning rules, however, suffer from being slow and inefficient. They rely on a random shuffling to find the proper connections, a little like shuffling and reshuffling a random deck of cards to try to arrange the cards by their suit.

In 1985, however, a new learning scheme was developed by Rumelhart, Hinton, and Ronald Williams, and independently by Dave Parker and Paul J. Werbos. Called *back propagation,* it is a breakthrough in training multilayered neural nets. In back propagation, the network is given not just reinforcement as to how it is doing on a task; going one step further, information about errors is filtered *back through* the system and is used to adjust the connections between the layers of neurons to make a better performance.

Back propagation is supervised learning, and it has proven much faster than other learning schemes in training hidden units in a multilayered neural net. The back propagation rule is not without its own problems; it is doubtful that neurons in real brains use this kind of feedback learning. But there is evidence that the brain uses enormous feedback among its neurons, and many connectionists think that more biologically plausible forms of the learning rule will be found.

Back propagation also presents a paradox for brain and mind researchers: The neural nets are trained by being given feedback on how their output compares with the correct answers, but to do that, a neural network must have some knowledge of what is and is not a correct output. In back propagation, that knowledge is provided by the programmer. For neural networks adequately to model the brain, however, there must be some way for the networks themselves to know the correct answer.

One possibility, suggests psychologist John Skoyles of the University College of London is that the brain trains itself "bootstrap" style, using one network to tutor another. For example, children learning to read aloud may first learn a few simple recognizable words, generally on a trial-and-error basis. The knowledge of how these few words are pronounced can then be used to train another network in the brain to pronounce smaller groups of letters present in those words. The knowledge of how to pronounce groups of letters can then be used to train yet another network to use the more sophisticated reading abilities of adults, which rely solely neither on recognizing whole words nor sounding out letters. Climbing a cognitive ladder, the brain trains a series of networks of increasing sophistication and boosts itself to a higher level of information processing.

The back propagation learning procedure has spawned an

explosion of interest in experimenting with multilayered neural nets. In one study, Geoff Hinton used back propagation to teach a neural net about the relationships between the members of two families. Hinton used two hypothetical family trees, one of an English family and one of an Italian family. (See the Figure below.)

Hinton trained his neural network simply by giving it examples of relationships within the family, such as "Sophia's mother is Lucia" and "James's wife is Victoria." Hinton gave the network one hundred examples of various relationships between members of both families. After training, the network could recall any of the relationships; given "Who is Sophia's mother?" for instance, the network would answer, "Lucia." The network could even fill in the details of relationships it hadn't been trained on by making inferences about the relationships it had seen.

More important, Hinton's network was able to pull out the relevant features necessary to learn about the family, even though it was not given the information. Although neural networks can make generalizations about groups of things that are outwardly similar, Hinton purposely made his network's task more difficult. The training information given the network was all in the same format—"Margaret's husband is Arthur"—and so there was noth-

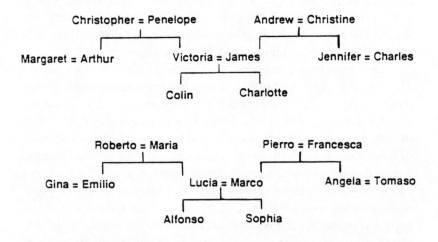

By being given examples of relationships among individuals in these two family trees, a neural network learned to distinguish the two families' different nationalities, the sex of some individuals, and the age differences among the three generations.

ing distinguishing about the various pairs of people and relationships between them. The only differences were in how the people were *related,* and the neural net had to make generalizations about the family trees solely from that information. During training, the hidden units drew out these relationships. Like the hidden units in the Boltzmann machine, various hidden units in Hinton's net took on representations of the world, capturing important features such as age, nationality, branch of family tree, that the father of a middle-aged person is an older person and that the mother of an Italian person is an Italian person.

Hinton's family-tree neural net learned only what it needed to make its decisions. In the example the network was trained on there is no mention of age, sex, or nationality. But because in these family trees the Italians are related only to other Italians, and the English only to other English, Hinton's network pulls out this English/Italian distinction to do the task. The network doesn't have a particular label for them, but there is a hidden unit that turns on for all the Italians and another that turns on for all the English. The network produces its own categories that it needs to in order to do the task.

For example, when Hinton's network is given the relationships between members of the families—"Jennifer's mother is Christine"—it doesn't need to know the sex of the first person. Jennifer and James can have the same mother, regardless of their sex. But the network does need to know the sex of the person at the other end of the relationship: Jennifer and James's mother must be a woman. The hidden units in Hinton's network reflect this fact. They don't capture the sex of the first individual in a relationship because the network doesn't need to know the sex of James to answer a question like "Who is James's mother?" But the sex of the second person, James's mother, *is* represented. The network has categories for mothers and fathers, sisters and brothers, uncles and aunts, grandmothers and grandfathers, and nieces and nephews, and they are all sex-linked. The network makes the distinctions it needs in order to get a right answer, but nothing else. The back propagation rule enables Hinton's network to learn to make categories about the people in the two families, but it isn't necessary that specific hidden units do specific tasks. The *pattern of activation* across several of the units would still distinguish the nationality, generation, and the rest.

Whatever their impact on understanding the brain and mind, neural nets may go far beyond their use as theoretical models of brain circuits and cognition. Certainly, the new learning abilities of neural networks are of interest to cognitive scientists. But the research of neural networkers is downright Promethean to a group of researchers who for the most part don't particularly care about neurons, brains, or minds.

These are the machinemakers. The startling news that connectionists passed down to engineers was that it might be possible to make neural networks, which had existed mostly as theoretical simulations in the computers of academia, into real "live" thinking machines.

CHAPTER SEVEN

MACHINE DREAMS

Putting Neural Nets to Work

Let's face it, behind part of the interest in connectionism
is that dirty little secret that research in nuclear physics had
during the thirties—that maybe you can *build* something with it.
—Gary Lynch

It doesn't look like much—a chessboard-sized chunk of wood
festooned with wires, switches, and assorted electronic gizmos.
Still, what the circuitboard lacks in appearance is made up in
historical significance, for it represents one of the first attempts to
hard-wire neural network theory into practice.

The board, a "demonstration model" of Hopfield's theoretical
neural network, was built by John Lambe, a solid-state physicist at
NASA's Jet Propulsion Laboratory in Pasadena, California. Lambe
realized early that Hopfield's theories might lead to the creation of
a radically new type of thinking machine. The switches on the
board can be turned on or off in different patterns to program
various memories into the network. If the network is then presented with a new input pattern, circuits will retrieve the stored
memory that most closely matches that input. The board's ability
to hold and retrieve memories is the first concrete demonstration
that Hopfield's mathematical theories on neural networks can be
used to create a working machine.

The Neural Net Industry

That simple demonstration model was quickly superseded by a
fledgling industry in neural networks. The First Annual Interna-
tional Conference on Neural Networks, held in San Diego in 1987,

drew thousands of participants representing every academic size, stripe, and creed. Filling auditoriums, poster sessions, and display booths, engineers and brain scientists alike traded optimistic projections about the future of neural net computers. Indeed, the poster announcing the neural net conference made no pretensions about how the organizers viewed the potential of neural networks as thinking machines. At the top of the broadside was the clarion call: "Come Join the Dawning of the New Era." Neural nets had hit the big time.

What was formerly a quiet, almost covert academic study was now one of the hottest new areas for technological investment. Neural networking has captured the imagination—and pocketbooks —of many, and may become the "genetic engineering" of the next decade. The idea that neural networks could be used to create a new kind of computer gathered momentum as news of Hopfield's research spread throughout the scientific community. Though other scientists have developed neural network models as research tools in psychology and neuroscience, Hopfield's work was influential in a different way. Hopfield explained the remarkable computational properties of these networks in the language of physics and mathematics, the language engineers and computer scientists understand best.

"The thing about Hopfield's models, and models like it, is their simplicity," says Satish Khanna, an engineer who designs neural network chips at NASA's Jet Propulsion Laboratory. "Many people had drawn pictures with sets of neurons on one side and the other, and connections between them, but the whole thing was a jungle. Hopfield drew essentially the same diagram, but did it in such a way that it allowed one to look at the whole picture as a much simpler architecture—and then it was possible to actually start thinking about putting these things into electronic components."

Hopfield also did some lobbying on behalf of connectionism. Because he was a well-known physicist with vast experience in working with electronic computer circuits, his reputation in the scientific community helped bring neural nets to the attention of machinemakers everywhere. A new group of physicists, engineers, and computer designers started paying attention to neural networks, as did newspaper and magazine reporters. "What happened is interesting from the point of view of the sociology of science," says Brown University's Jim Anderson. "Neural networks have actually been around a long while, but Hopfield is a big-name physicist at Cal Tech—and big-name physicists have a media access that you wouldn't believe. So all of a sudden neural networks started getting a lot of attention—it's been the greatest thing that could ever happen to the field."

Hopfield's research was just the thing that computer designers had been waiting to hear. Over the past two decades, conventional computer chips have approached closer and closer to the technological limits of their design, but they are still woefully inadequate at doing brainlike chores like recognizing patterns and making generalizations. Hopfield's message was simple: Engineers might not be particularly interested in how the brain works, but if they wanted to design a new kind of computer that does the useful tasks the brain can do, they might try modeling it after the brain itself.

A wave of neural net research is now spreading through laboratories across the country. Scientists at AT&T, TRW, Texas Instruments, IBM, General Electric, and NASA's Jet Propulsion Laboratory are all scrambling to find the best way to put a neural network on a silicon chip. For example, Lawrence Jackel and his colleagues at AT&T Bell Laboratories, where Hopfield has a joint appointment, are testing a chip that has 512 neurons. A Japanese firm, NHK Science and Technology Laboratories, is working on a neural network that recognizes typewritten characters, even if they are distorted. SAIC Technology Research in San Diego offers a $15,000 circuitboard that can be added to a conventional computer to simulate a neural network. It can update the connections between neurons at the rate of two million per second. Similar add-on boards are quickly appearing. TRW offers a commercial version of a neural net simulator similar to the one they originally developed for the Pentagon's Defense Advanced Research Projects (DARPA).

The Department of Defense, ever mindful of the applications of thinking machines for war machines, announced in 1988 that it plans to spend as much as $400 million over the next eight years supporting a wide range of neural net projects—one of the largest research efforts it has ever undertaken. In one possible defense application, Paul Gorman, at Allied-Signal, with help from Terry Sejnowski, has trained a neural net to recognize underwater targets from the objects' reflections of sonar signals. After only three hours of training, the neural network was able to outperform humans and a conventional computer program that took more than ten months to design. In another project, the University of Pennsylvania's Nabil Farhat has built an optical neural net that stores how radar signals reflect from various types of vehicles. It can match the correct one from as little as 10 percent of the complete radar pattern.

This wave of research is not confined just to the big chipmaking companies. According to Edward Rosenfeld, Editor of *Intelligence,* the first newsletter devoted to neural net technology, more than two hundred companies hoping to exploit neural nets commercially have started up over the past two years, and the number is growing. One such company is Nestor, Inc., a company started by Brown University researchers Leon Cooper and Charles Elbaum. Their program, NestorWriter, employs a neural net type of design and can be run on an IBM home computer. It reads handwritten characters written on a pen-sensitive pad. Nestor also has a network-based program that can recognize any of the 2,500 handwritten characters used to write Japanese.

Because neural networks are adept at making generalizations from a set of complex data, they are often used as analytical tools. One financial services company in Irvine, California, for example, is using a neural network to judge the worthiness of applications for loans. According to one study that compared the performance of the neural network to a computer program the company had previously used to evaluate loans, profits would have been 27 percent higher had they used the neural network. In another project Alan Lapedes, a physicist at Los Alamos National Laboratory, is using a neural net to determine whether particular strands of DNA are responsible for the production of particular proteins. The network makes the right prediction about 80 percent of the time, which is better than the statistical methods previously used. Lapedes's network even uncovered an error in a library of information on DNA sequences collected by GenBank, a research center run by Los Alamos.

Another researcher, Michael Kuperstein of Neurogen in Brookline, Massachusetts, uses a neural net to give a robot rudimentary hand-eye coordination. His robot uses two cameras for eyes and has a double-jointed arm that can grasp objects in front of it. Comparing signals about the object's position coming from each eye, the robot eventually learns by trial and error how to grasp an object placed anywhere in front of it. His work is being sponsored in part by NASA, which is interested in creating robots to help assemble parts of a space station in orbit.

The Matchmaker

One area where neural networks might soon find gainful employment is in pattern-matching. While making an exact match between an input pattern and one stored in memory is a relatively easy computational task, finding the *closest* match among many patterns not quite alike requires much more computation. One practical application of such matching tasks is in data transmission. A typical black-and-white television image is made of hundreds of thousands of dots, called pixels, that have intensities ranging from white to gray to black. Normally, when the image is transmitted, a string of data representing the intensity of each pixel is sent. Researchers have long been searching for ways of compressing that huge amount of data into a smaller string of information that when "decoded" will still produce an image that looks almost as good as the original.

One proposed method to do this is to create a hypothetical thousand-page "book," with each "page" bearing a slightly different four-by-four pattern of pixels. The pixels in a television image divided into a grid of four-by-four patterns, and each pattern in the grid is matched with a pattern in the thousand-page book. Then, instead of transmitting all the pixels in the image, a string of *page numbers* representing each four-by-four pattern is transmitted, drastically reducing the amount of data sent. At the receiving end, the image is recreated by looking up the page numbers in an identical book and putting those patterns in place. The resulting image won't be exactly the same as the original, but it will be close enough for most television needs. The technique can also reduce the amount of data transmitted over phone lines. Despite its money-saving potential, however, this data-compression technique has not been put to much use. "People have known about this trick for years and years," says Hopfield, "but no one has done it, because the 'close-match' problem is too difficult and the rate that television pictures are transmitted is too fast."

But neural networks are ideal for these applications because they store and retrieve memories differently than conventional computers. In a computer, each memory has what's called an address, analogous to the addresses on houses on a street. To retrieve the memory, the computer goes to the address, rings the doorbell, and the memory comes out. Like a house address, the computer's memory address is just a number; it has no connection to the actual content of the memory. To find the phone number that matches a particular name, the computer takes the input name and goes down the street, ringing doorbell after doorbell and comparing it to the memories at each address, one by one, until it finds a match.

Though this memory-retrieval system is cumbersome, conventional computers operate so quickly that they can perform very sophisticated matching tasks. The word-processing software with which this book was written, for example, has a dictionary that will flag misspellings and suggest a word, with the correct spelling, that I might use. It works quite well, offering *separate* when I type *seperate*.

But sometimes the program runs into dead ends. When it comes across a simple misspelling such as *tommorow,* for instance, it's baffled. It doesn't know to replace *tommorow* with *tomorrow,* because it's not programmed to deal with that particular kind of misspelling. A conventional computer has to be explicitly programmed to look at a name and give you a number, look at a number and give you a name, look at a last name and give you a first name, and so on. For every aspect of a computer's memory-retrieving ability, a separate set of instructions must be written.

A neural network, however, doesn't need separate instructions for retrieving data, because the storage of memory and the ability to recall it from any of its parts are both part of the same system. Consider the hypothetical neural network discussed earlier, which consisted of a thousand students turning switches on and off. Suppose you give the students identification numbers, and tell them that if their switch is on, they should hold up a big sign that has a 1 on it. If the switch is off, the student should hold a sign with a 0 on it.

In a conventional computer, information is encoded by using strings of 1's and 0's; the letter A, for example, can be represented as a series of 1's and 0's such as 00001, the letter B as 00010, and so on, up to 11010 for the letter Z. In the same way, letters can be encoded in the student neural net computer. Employing the coding format used with ordinary computers, a thousand-student neural net is able to produce a memory with about two hundred type-written characters. On a neural net, these strings of 1's and 0's are not stored as a set of switches in a separate memory bank, as they are in a conventional computer; instead, the resistors in the wires connecting all the students are adjusted so that the network has a variety of stable "resting" states, each state producing the 1's and 0's corresponding to a different memory containing a different string of information.

For example, the various strings of information might be your friends' names and phone numbers. You'd construct one stable state that produces, say, the 1's and 0's that encode WILLIAM F. ALLMAN = 202 965 5555. Another stable state might produce the code for SCOTT FERGUSON = 314 747 5555. You can continue to add names and numbers, depending on the size of your neural net. When you've finished entering the information, you have a neural net phone directory.

To use your phone directory, you have only to walk through the gym and clamp down the switches on those students who represent the various letters of the name WILLIAM ALLMAN. Then you tell all the other students to watch their meters and flick their switches on or off. The student neural net will eventually come to rest in a stable state, and since the only stable state of the neural net that includes the 1's and 0's for WILLIAM ALLMAN also includes the 1's and 0's for the number 202 965 5555, the network produces the correct phone number.

The directory will work in reverse, too. If you clamp down the student switches that correspond to 202 965 5555, the system will settle into that same stable configuration, and you can read the name WILLIAM ALLMAN. Or suppose that all you can remem-ber is that you know someone named William who lives in Wash-ington, D.C., and you want his last name and number. You clamp down the switches corresponding to WILLIAM and the area code for Washington, 202, and the student neural net will evolve to the rest of the stable state and produce the rest of the memory.

If your memory is *really* selective and all you can remember is W–L–I–M A–L–A– = 2–2 9–5 5–5–, the student neural net will still be able to complete the pattern and display the rest of the information. In fact, Hopfield has shown that clamping as little as

5 percent of a name or number into the net will produce the rest of the memory. The system can also deal with slightly incorrect information, making the best match of the faulty input with one of its memories. BILL or BOB ALLMAN, for example, still contain enough information to summon the correct telephone number as long as there are no other similar names.

This kind of "associative" memory retrieval, like many aspects of neural networks, is very similar to how our brains operate. In a typical computer, the more information you add to a memory, the longer it takes for the computer to make the one-by-one comparisons it needs to retrieve it. But on neural networks, the more information you put into a memory, the more stable that memory is, and the easier it is to retrieve. The same is true for brains: Adding additional information to a memory has long been a trick used by people who want to improve their recall. People usually find that it's easier to remember that someone *is* a cook than that they are *named* Cooke, so a common mnemonic trick is to associate the name of a person you've met for the first time with some characteristic about him or her. If a woman is named Taylor, you might make a mental note that Ms. Taylor has an outfit that is well "tailored." Or try to imagine that Mr. Fischer's face looks like a fish. Even if you can't think of an appropriate pun for a name—say the name is Magilicutty—adding an extra bit of information such as a nonsense rhyme—"Magilicutty, made of putty" —often helps you to remember it. Even though adding such nonsense to a person's name does nothing to increase the meaning of the memory, it makes the memory itself stronger.

Another brainlike feature of neural networks is their ability to make a "best guess." When you listen to someone speaking English with a foreign accent, your brain tries to match the unfamiliar sounds to familiar words in your memory. Even if the accent is very heavy, your brain will broaden its range of possible matches to make a guess about the sound. On a neural net, this is achieved by adjusting the network's threshold level. Sometimes there are "flat spots" between the valleys representing memories in a neural network. When an input pattern falls into one of these flat spots, it doesn't go anywhere; the network says, "I don't understand what's been said." A researcher can adjust the threshold by increasing the width of the valleys throughout the network; when an input is far from a "true" match, it will fall into a valley anyway, producing a "best guess."

Neural Nets in Space

Because neural nets store memories among groups of neurons, data stored on a neural net chip is potentially simpler, hence more reliable, than conventional memory devices like tape recorders or computer chips. When a memory is stored, its information must be put into the device physically. In computer chips, electrons are moved from one point to another to construct a memory. It's very easy to move electrons around in a silicon chip, but it's very easy to knock a few out of place, too, destroying that information. Neural nets store information not as a single blip of electrons but as a pattern of activation among all the neurons. It takes much more energy to destroy a pattern than to knock out a few electrons, so a neural network is less susceptible to losing its memory.

Neural nets' resilient memory storage makes them ideal for applications in machines that must operate in hard-to-reach places on Earth or in outer space. Researchers at NASA's Jet Propulsion Laboratory (JPL) are studying the use of neural networks in their planetary probes, to protect critical information from damage due to cosmic rays. Researchers would like to make on-board memory storage devices as small as possible to save weight, but the more compact a memory device, the greater the chance that a cosmic ray will hit a vital circuit. A neural network stores information more securely because, like a brain, it can lose as much as 5 percent of its neurons and continue to function nearly as well. "We lose five percent of our neurons every year," says Hopfield. "Yet our mental capacities don't diminish—in some cases they improve. But if you cut five percent of the wires in a conventional computer, it grinds to a halt."

Since neural networks are good at pattern matching, they might also be useful on automatic rovers that roam the surface of a planet, says the Jet Propulsion Lab's Khanna, the director of the neural net development project there. One problem in creating such autonomous vehicles is determining how it will be steered. Controlling the vehicle by electronic signals sent from Earth would be impossible because such signals can take as much as a half hour to travel between Earth and Mars. If an Earth-based technician was monitoring the rover's travels via an on-board television camera and suddenly saw that the vehicle was about to roll into a deep crevasse, his signal would arrive too late.

Because of these signal delay problems, JPL researchers would like to give such a rover the ability to recognize basic landforms by itself. Neural nets hold promise for doing such recognition tasks because they can categorize a new pattern as part of a class of patterns already stored. In one experiment by McClelland and Rumelhart, a neural net was trained to simulate the experience of a little girl who encountered several different types of dogs in her neighborhood. Suppose the girl were told that the different animals have different names, such as Fido and Rover, and that they were all examples of *dogs*. In the experiment, the dogs were represented by a string of numbers in the neural net; Fido could be represented by a string such as 101–1100000; Rover, 011–1100000.

In much the same way that dogs in the real world share many attributes, the number strings that represented different breeds of imaginary dogs were all slightly different variations of a string of numbers that represented a "prototype" dog. After the neural network was shown various examples of these imaginary dogs, it could recognize any new dog pattern as part of the overall dog category. With further training, the network was able to distinguish between dogs, cats, and bagels.

This same categorization ability might enable a Mars rover to recognize mountains and ravines. A neural net with three stable memories, each corresponding to a particular type of landform, such as mountains, ravines, or boulders, would produce a "closest match" between any new pattern that slightly resembled it.

Such a space-age neural net, however, is still the stuff of dreams. For the present, Khanna and a colleague at JPL, Alex Moopenn, are trying simply to build a crude associative memory into a silicon chip. The researchers manufacture a tiny grid of horizontal wires—the distance between each wire is about a hundred microns—then lay a grid of vertical wires on top of it. Sandwiched between the crisscrossing grids is what Khanna calls magic material, hydrogenated amorphous silicon. The material is similar to what solar cells are made of, and it has an important property; when a low current is applied to it, it displays what electronic engineers call a high resistance—it won't let current through. But if a high current is applied to it, it will suddenly, and permanently, change from high resistance to low resistance.

In a way, you could think of it as a "fuse" in reverse. Ordinary household fuses allow current to pass through them as long as the current is low. But if the current is too high, the fuses trip, and the current is stopped. Khanna's magic material works in the opposite way; the material is normally reluctant to let current pass through, but when it gets a high current, it changes its electronic structure, opening the floodgates and letting the current through.

The researchers program memories into a neural net chip by first laying a horizontal grid of wires over the vertical grid, with a thin layer of "magic material" sandwiched between to prevent current from flowing between the two grids of wires. When the researchers apply a strong positive current to one wire in the horizontal grid and a strong negative current to a wire in the vertical grid, the material between works its magic. Current from one wire alone isn't strong enough to change the resistance in the material, but at the point where the two wires cross, the *combined* current of the two wires is strong enough to cause the material to lower its resistance. That creates a tiny, permanent electronic "hole," allowing the current to flow from one wire to the other.

By selectively applying high current to wires in both grids, Khanna and Moopenn can custom-design the network's connections and create various stable states. Most important, they can enter information without having to use costly transistors or other complicated silicon components that take up a lot of space on conventional computer chips. "It's a question of efficiency," says Moopenn. "If you wanted to store the same amount of information in standard digital computers, you would need at least two transistors to store each bit. That is almost a hundred times more expensive in terms of space and cost than our neural net. The basic difference is that we are storing the information *passively*—it's stored in these connections, not in active electric components like transistors."

Khanna and Moopen hope to find a method temporarily to open and close the holes in the silicon, instead of their being set permanently, and make the resistance in the material variable instead of on-off. If a neural network has electrical shades of gray, it can represent a broader range of values than 1 or 0 and therefore perform more sophisticated tasks.

The projects of Khanna and other researchers around the country are still at an early stage of experimentation. Scientists aren't sure which design is best for a neural network chip, so they are trying all kinds of architectures, from variations of Hopfield's original model to decades-old designs recently rediscovered in the rush to put neural nets in silicon.

Computing with Light

Others are using neural networks in computers that operate not on electricity but *light*. Cal Tech's Demetri Psaltis is rejuvenating this small but growing corner of research known as *optical computers*. Light has been used to shuttle information around for several years; long-distance phone calls, for example, are often carried by pulses of light speeding through optical fibers. On a compact disk, light is turned into near-perfect musical recordings. In optical computers, light is the current that flows through the machines.

The idea of computing with light is not new, but only now are optical computers, beginning to be taken seriously. "I first got into optical computers about fifteen years ago," says Psaltis. "By then they had already gone through a peak cycle of interest and were sort of down and out. I remember going to conferences where only five people attended—and those were the researchers presenting papers. But now people are starting to see the limitations of conventional silicon computers and accepting the idea that some major change is needed rather than just some improvements along the basic paths. So optical computers are again being looked at seriously." Light beams are ideal for neural net computers because they need no wires to move from one point to millions of other points and can cross one another's paths without interfering. "Communication is what gives a neural net its power," says Psaltis. "And optics is a beautiful technology for communications."

Psaltis has a demonstration model of his optical neural net set up in the basement below his office. The darkened room is filled by a large table laden with instruments that looks a little like a setup for a toy train or slot-car racers. But in this race, the horses are photons of light; tiny mirrors are arranged in an oval track, and the eerie light of a laser squeezes out of a box at one end and zips around the table. At another point in the oval is a camera connected to a television monitor. Displayed on the monitor are large letters: CAL TECH OPTICAL NEURAL COMPUTER.

If you put your hand between two mirrors, partially blocking the beams of light, the words OPTICAL NEURAL COMPUTER disappear from the screen. But slowly, like a Polaroid picture developing, the letters come back, even though your hand is still between the mirrors. The letters are a little fuzzier, but they are there. No matter where you put your hand to block the light, the image will dim for a minute, then return again.

This is a visual version of an associative neural net. Even though half the memory is blocked out of the network when a hand is dipped into the light beams, what's left is enough to reconstruct the rest of the image. No matter which part of the pattern is taken out, the neural network completes the rest.

It looks like magic, but it's not all done with mirrors. The core of an optical neural net is a *hologram*. Holograms are perhaps best known as three-dimensional images created in two-dimensional pieces of plastic. An object normally appears three dimensional to us because when light hits its surface, the light bounces off in all directions. When these light rays bounce off an object and collide with a conventional film negative at the back of a camera, they leave a trace of their brightness as they passed through. The film, in other words, records the *intensity* of the light beam. When the film is developed, that intensity is reproduced as a two-dimensional photograph of the object.

When light passes through holographic film, the film records not only the intensity of the photons in the light beam but also, like buckshot passing through a wall of Jell-O, the *direction* of each photon when it hits the film. As the light tunnels through the hologram, it changes the electrical structure of the holographic material; this changes how the hologram subsequently bends another light beam when it passes through.

Once an image is recorded on the hologram, the hologram behaves a little like a mirror; light comes at it from one direction and is shunted off in another direction. But while a glass mirror simply reflects all the light in the same direction, a hologram is actually more like an array of *billions* of microscopic mirrors, all set at varying angles of deflection. When ordinary light is shone on a hologram, the light passes through and is deflected in many different directions, recreating the scattering of light formed by the original three-dimensional image.

The same principle used to create three-dimensional images can be used to construct the communication links in a neural net.

Since light beams come into a hologram and are redistributed to other points, each part of a hologram can be used to connect a neuron to another neuron. Since light waves don't need wires, optical neural nets can contain a huge number of connections in a tiny space.

Because holograms can store huge amounts of data, optical neural nets are ideal for what Psaltis calls random problems. Suppose that a police force wanted to use a computer to match a set of fingerprints with one of the millions of prints in their files. Someone programming a conventional computer to do the task would look for a rule that distinguishes one fingerprint from the others. But in this case, no one knows what it is that makes one set of fingerprints different from another; fingerprints are different for "random" reasons, so there is no way to specify that randomness in a rulelike way. The computer must therefore have all the fingerprints stored in its memory and sift through all until it finds a match.

Since neural networks fuse the operations of memory storage and retrieval, an optical neural network has the potential to store millions of patterns and instantly sift through them. In one experiment, Psaltis created an optical neural net that held the images of people's faces. When the network was given an input image that was just half of one of the faces, the half-image went round and round the oval track, reinforcing itself with each pass. Eventually, the correct image of the rest of the face appeared.

Engineering the Brain

In another project at Cal Tech, researchers are taking a different approach to creating a neural net in a machine, this time by building small electronic circuits that closely mimic the actual form and function of circuits in the brain. At a lab near where Psaltis works, Michelle Mahowald, a graduate student, places a pinwheel above a tiny piece of silicon nestled in a bed of wires. She flicks a switch, and the pinwheel begins to turn slowly. On a nearby video monitor, a gray pinwheel image appears, whirling. The video monitor is connected to the tiny chip. The chip is telling us what it sees.

This tiny bit of silicon can visually detect motion better than even the largest computers, says Mahowald. But there is no computer connected to it at all. The chip is a silicon retina—perhaps the first step toward giving robots neural-net sight.

The silicon retina was made by one of the pioneers of chipmaking technology, Cal Tech's Carver Mead. Mead, a long-time associate of Hopfield's, was the first person to put a Hopfield net on a real silicon chip. Using standard chipmaking technology, Mead constructed a twenty-two-neuron design that used transistors to simulate the neurons and the connections between them. When Mead began to design his silicon retina, however, he took a slightly different tack, looking to a machine that had already solved some of the problems that he was trying to tackle—the brain. "The most horrible misconception today is that neural computation is kind of clumsy, slow, and inefficient, and that we can really do much better with our digital computers," he says. "That is far from the truth, and it obscures a very deep piece of awareness: The brain is an amazing computational instrument—awesome by any standards, and awesome by any standards way beyond what we can imagine today."

Mead's retina is patterned directly after the retina in the eye of a mammal because the mammalian retina is actually simpler than the retina of primitive vertebrates. A frog has a tiny brain, but its retina is quite sophisticated; the retina has a built-in "bug detector" that senses when small dark irregular spots are moving relative to a stationary background. So-called higher animals like humans have more-sophisticated brains; many of these kinds of bug-detecting tasks are done by the brain itself, so their retinas are somewhat simpler.

"Here's the chip," says Mead, tossing a small metal-and-plastic rectangular one on the table at the restaurant where he is having coffee. "You just put a lens on it, and it will 'see'—it can compute how objects are moving. Of course, that's only *one* of the things your real retina does, but it is an important thing. And it's one of the things you can't do with conventional computer vision systems. *You just can't do it.* There are people putting supercomputers behind television cameras to try to do stuff like this little chip does, and it doesn't work. That's why I did the motion first—it's something they can't come close to."

The chip was built by the normal silicon process for making conventional computer chips; the only thing different about the chip is its design. Mead made it with conventional technology because he didn't want to compensate for a poor design with extremely precise components. Compared to electronic circuitry, real neurons are extremely sloppy and error-prone, and Mead wanted his chips to reflect this brainlike characteristic. "Neurons are nasty, noisy, imprecise things," says Mead. "So I'd be kidding myself if I made the components in my chip accurate to a tenth of a percent or something like that."

Mead's artificial retina uses a bank of photoreceptors that detect the brightness of light. As in a real retina, the silicon retina's receptors are connected so that they take the *logarithm* of the intensity of light. To reprise your high-school math, logarithms are a way to multiply two numbers by using addition: You take the log of each number, add them, then look in a table to find which number has that log.

According to Mead, expressing the intensity of light in logarithms is a brilliant engineering decision for nature to have made. When you look around your living room at light and dark objects, your retina sees the difference in the intensity of the light on those objects and expresses that difference as a logarithm. If you turn off most of the lamps in the room, the intensity of the light in the room may be ten times lower than it was before. But because your retina uses logarithms, the difference in intensity between light and dark objects in the room will still be the same, and the objects will look the same relative to one another.

The ability to make out the *difference* in intensity of light is what makes it possible for us to avoid bumping into chairs under different lighting conditions. But we sacrifice something; though we're good at judging the difference in light intensity between two objects, we're not especially good at judging the *absolute* intensity of light. Sitting and talking on the back porch on a summer's evening, you may be suddenly surprised to find that it has become dark. Because the intensity of light was slowly diminishing, but the *relative* brightness of everything stayed the same, you didn't notice it.

Mead has joined neurobiologist Gary Lynch and microchip expert Federico Faggin in a neural net company called Synaptics. Based in San Jose, California, Synaptics, like many other recently established companies, hopes to use neural network technology to solve some of the problems that have defied solution by more conventional means. One such problem is getting a computer to understand human speech. If there is a massive stumbling block to integrating computers into our society, it's that there is no easy way to communicate with them. A computer may be one of the fastest in the world, but to communicate with it requires a keyboard, the basic design of which hasn't changed much since the early 1900s.

Computers That Listen

Despite years of research, and more than a few premature announcements that it was "just around the corner," true speech understanding has eluded the most talented artificial intelligence researchers working with the most sophisticated computers. Forget understanding meaning and structure. Computers still can't reliably tell the difference between a spoken "*b*" and a "*d*."

There has been some progress, however. Many systems respond to specific words spoken by specific people. Texas Instruments, a pioneer in speech recognition research, used the fact that every person's voice has a unique pattern to create a computerized voice-controlled entry to one of their laboratories. A researcher would stand inside a small room and repeat several words into a microphone. The computer matched the researcher's voice to a set of sample words that the researcher had stored earlier. If the sound matched, the door was opened. The system worked adequately, but occasionally someone would have trouble getting in the laboratory because her voice was slightly changed by a cold.

Problems such as the variability of different people's voices and sensitivity to background noise are endemic to conventional methods of speech recognition. "The whole AI approach to understanding speech is a dead end," says Bruce Ryon of Votan, another neural net startup company. "It's built on a house of cards. Extraneous sounds such as a squeaking chair just drives these systems nuts. But you and I can have a conversation and understand each other even if we are standing near a roaring jet or are in a roomful of people talking." The ability to understand at least some aspects of speech is found even in animals not usually known for their brilliant cognitive processing—quail. A project at the University of Texas at Austin and Arizona State University found that the lowly quail was capable of doing what the most sophisticated computers cannot: telling the difference between *b* and *d*. The birds were trained to peck at a lever on a food dispenser when they heard the words beginning with *b*—responding to *bog,* for example, but not *dog*—or vice versa.

That a tiny quail brain can outperform a computer at such rudimentary recognition of sounds suggests that neural networks might someday make inroads to solving speech perception. That's just what many of the neural net companies, both in the United States and in Japan, are banking on. One Japanese company, Asahi Chemical, is reportedly negotiating with neural net researcher Teuvo Kohonen, of Helsinki University, to create a speech recognizer. Synaptics hopes to use Carver Mead's artificial version of a cochlea, part of the ear, to understand speech. Both Hecht-Nielsen Neurocomputer and Votan are working on voice-recognition systems. Votan is test-marketing a neural net system that can recognize numbers spoken over the telephone, an ideal application for credit card companies and banks. It's actually even better for banks when the common language is Spanish, says Ryon, because the Spanish words for different numerals are longer than the English, making them easier to recognize. The Votan system is being tested by a bank in Buenos Aires.

The New Engineering

According to Mead, the "dawning of the new era" of neural networks will have two effects. First, it will push neuroscientists toward trying to understand the brain as a computing machine. "We simply have to do our homework, figure out what the *biological* system is doing first," he says. "Once we've done that, we can build a system, and compare it to the real system, and ask the hard questions."

The other effect, says Mead, is that a new kind of engineering will arise. The old engineering has focused on using the technology that was most advanced and easiest for quick gains. But researchers are beginning to realize that many of their most difficult engineering problems have already been solved—by the mass of neurons inside their skulls. "We have a machine to study, the brain, that is the result of millions of years of evolution," says Mead. "What's more, we have the result of these millions of years of evolution to study it *with*. So there's no point to try to use trial and error to recreate something that we have sitting in front of us. We are in no better position to copy biological nervous systems than we are to create a flying machine with feathers and flapping wings. But we can use the organizing principles as a basis for our silicon systems in the same way that a soaring bird is an excellent model of a glider."

The neural network meeting in San Diego generated mixed feelings among connectionists. The conference heralded the news that neural networks were being taken seriously as a new design for practical thinking machines, but it also gave notice that the field wouldn't be as soft-spoken as before. Money may be made with neural nets, and along with the rush of commercial interest comes the inevitable overselling of the new machines' potential. In fact, some researchers worry that the sudden influx of startup money and machines—and the accompanying hyperbole and press releases—may backfire. "Some of these people are selling empty promises," says one concerned connectionist. "They're saying, 'To hell with the future; let's get rich off the present.' "

Though some neural net researchers have made extravagant claims about "building a brain," most know that there is a lot of work to do before neural net machines start performing practical tasks. "No one is even *remotely* contemplating doing anything *remotely* like a brain," says Mead. "We're just trying to do little pieces—an artificial retina or cochlea. If we can do just these little pieces, we'd be very happy. It's a start. In another ten or twenty years we can go a long way."

While research on neural networks may enable engineers to build new mindlike machines, the impact of connectionism may extend even further. In the same way that the new model of the mind helps cognitive scientists explore how groups of neurons interact in the brain to produce the mind, neural networks may be useful in examining areas where groups of brains interact with other brains to produce language, society, and culture. A few researchers in linguistics, anthropology, and sociology have already begun to apply neural nets to these areas. Though their work is still at a very early stage, it appears that as in neuroscience and psychology, neural networks may lead researchers in these fields to reevaluate some of the cherished doctrines of the past.

CHAPTER EIGHT

WARRIORS OF THOUGHT

Winning the Hearts of Mind Researchers

A new scientific truth does not triumph by convincing its opponents, but rather because its opponents die, and a new generation grows up that is familiar with it.

—Max Planck

If there has been a scientific revolution in the study of mind based on ideas of symbolic computation, the counterrevolution is coming into view.

—James G. Greeno, editor of the journal
Cognitive Science

Jerry Fodor is on a roll. A big man, his round face capped with an incongruous dome of slate-gray hair cut in the bangs-down style of a schoolboy, Fodor is as charming, disarming, and congenial a scholar as can be found. At the moment, he is tearing the intellectual core of the connectionist movement to shreds—right in front of the people who helped construct it.

"The fact that there are people who have given up good theories for bad reasons I'll admit," he says to the packed room of connectionists at UCSD. "The fact that people have given up their theories when they ought to have stuck to them and seen their way out of their problems I'll admit, too.

"But that is simply irrelevant. *Your* theory has a problem that *my* theory doesn't. I don't think my theory has *any* problems that your theory doesn't. So by any reasonable criterion of how science should be conducted, you *gotta do something about your theory*. I'm suggesting throwing it out." Fodor hurls this last salvo with a smile. A few of the researchers in the room roll their eyes. Most of them, however, shift and squirm uncomfortably in their chairs.

The Empire Strikes Back

To a connectionist, listening to Jerry Fodor is a little like a wayward teenager listening to a lecture from his father. For the first time, connectionist ideas are being taken seriously by the establishment, but only to the extent that they are being called wrong. Fodor is one of the founding patriarchs of the "rules-and-symbols" approach to understanding the mind. First at MIT and now at City University of New York, he has helped shape the study of language and cognition for nearly two decades. Connectionism is a direct challenge to nearly everything Fodor believes about the nature of the brain and mind, and he isn't about to let that challenge go unanswered. So Fodor traveled to UCSD, the home of the new connectionist movement, to give an informal talk on why the researchers should see the folly of their ways and give it all up.

Before Fodor's arrival, the researchers in the PDP group tried to prepare for what they knew would be a difficult debate. They gathered together the weekend before Fodor's talk to discuss the points that he was likely to raise and how they should respond to them. But they were worried; David Rumelhart was out of town, and Don Norman, who heads the UCSD cognitive science department, would have to leave the talk early. Still, Patricia and Paul Churchland were going to be there, as were Francis Crick and the rest of the UCSD group. Terry Sejnowski was coming down from Cal Tech, where he was a visiting professor.

Given that the twenty or so people who assembled in the UCSD classroom that day for Fodor's lecture had all made career decisions to delve into connectionism, it might have seemed that the philosopher was walking into a lion's den. But that is just the kind of action Fodor craves. Like many of the scientists and philosophers in the old guard, Fodor relishes the thrill of debate, and he is one of the best.

Fodor's argument was as simple as it was technical. Basically, he was asking the question "What kind of cognitive machine is the mind?" Fodor answered that question with an assertion: Whatever kind of machine it is, one important characteristic is that if it is capable of understanding a proposition such as "John loves Mary," it is also capable of understanding a sentence such as "Mary loves John." In other words, it is capable of understanding the idea of *relationships* between things, whether it's Mary, John, or anybody else loving or anybody else being loved.

To most of us that sounds pretty simple, if not obvious. What's not obvious is how you can design a cognitive system capable of understanding that. In the traditional formalism that Fodor champions, the solution is easy. John and Mary are regarded as symbols, and their love is a relationship between the symbols. In Fodor's system, you can describe this relationship as a "sentence" in the "language of thought":

$$J \ (r) \ M$$

Where J stands for John, r for the love relationship, and M for Mary. Notice that it doesn't really matter whether J or M is in the first or third slot in the equation or whether the symbols refer to John or Mary; they could just as easily stand for Jack and Melissa or jam and marmalade. The r could represent *loves, hates, is the boss of, tastes like;* any relationship will do.

The ability to use symbolic sentences to represent everyday situations in the world gives the traditional symbolist approach its versatility and power. As Fodor terms it, this symbolist machine, like the mind, has *systematicity*. This systematicity is a *natural property* of any cognitive machine constructed with the classic symbol-and-rule approach.

Unfortunately for the connectionists sitting in the room that day, these kinds of systematic transformations are not natural properties of neural networks. A network representing the concept "John loves Mary" as a pattern among its neurons isn't able to deal automatically with the concept "Mary loves John." While a neural net may someday be designed to handle such relationships, no one as yet has discovered just how that might be accomplished. The connectionists knew that. Fodor knew it, too. He seized this point and wielded it like a cudgel. These kinds of systematic John-loves-Mary transformations are a natural property of real brains, Fodor noted, and naturally flow from the traditional formalist approach. Any model of the mind constructed with the classical architecture is *guaranteed* to be systematic. Further, since no other model of the mind—particularly connectionist models—carries that same guarantee, the classical architecture must be the correct representation of our cognitive processes.

To Fodor, this was all the reasoning needed to conclude that connectionism is a dead end. The connectionists in the room, however, weren't going to give up that easily. Joining in the debate early was Terry Sejnowski, who pointed out that in fact many connectionists were working on these relationships. Fodor was unappeased. "It's not hard to set up a network to represent these relationships," he retorted. "What's difficult to do is to give an architectural *guarantee* of systematicity. There is one and only one way to get guaranteed systematicity, and that is by using the classical architecture."

Francis Crick tried to counter Fodor's argument by suggesting that, in time, a neural network will be devised that will carry that guarantee. "Just because the model you have is adequate doesn't mean that it's the real one," he said. "The question is, What should be added to the PDP model to make it resemble the classical model? We will have to add something to do with relationships, something to do with sequence of events, probably something to do with attention. But it doesn't have to be *the same* as the classical model."

Patricia Churchland also joined in the debate, which began to grow more impassioned. She pointed out that Fodor's classical architecture has its own weaknesses, such as the difficulty in accounting for our ability to learn new concepts. In fact, said Churchland, it is precisely because the language-of-thought model has so many problems that the connectionists have been driven to abandon it for the neural network model, despite its inadequacies. "I don't mean to say that we don't need systematicity in PDP models," she said. "But just because we can't do that right now doesn't mean that your classical architecture is the only game in town. In the history of science there have been oodles of people who have said that their crummy little theories were the only game in town. And then, a hundred years later, you find that in fact there are lots of other games in town."

Fodor's reply was sharp. "You can only use those Kuhnian arguments *after* you've already won," he said, the pitch of his voice rising. "You can't use them as a way of exempting yourself from accounting for the facts. *I'm* the only game in town that can do systematicity. And given that, you've got to say either 'Systematicity isn't worth doing' or 'I can do systematicity, too' or 'There is no such thing as systematicity.' But you really have to say one of those three things. This is a scientific argument about a phenomenon everybody thinks a model of the mind ought to explain. One theory does, and the other one doesn't. The reply to that seems to be 'Yeah, but how do you *know* that the other theory won't *eventually* explain it?' That's not *polite,* somehow."

"But it takes *time* to build a good theory," shouted Paul Churchland, pounding the table.

Fodor shouted back: "It takes time to build a *bad* theory, too." His retort got a good laugh from the crowd and snapped the tension that had been building. Like lovers who suddenly realize during a quarrel that things have been blown out of proportion, the debaters grew quiet, and Fodor continued in a soft, almost apologetic voice. "Look, I am a philosopher by training and instinct, and I think that there is a deep metaphysical problem about the mind: How is rationality *mechanically* possible? And Turing had this wonderful idea—a remarkable, very beautiful idea. It was the first advance in thinking about the mind since Hume. What seems to me to be so dangerous about this PDP stuff is that it is on the way to convincing people to dispense with the *one* good idea that anybody has ever had in this field. That's why I feel so vehement about it."

The discussion went on a while longer and, like many such discussions, had helped sharpen the focus of the many problems still faced by the emerging new theory of the brain and mind. But also like many such discussions, once the main scientific points were made, the talk became more a game of rhetorical exchange.

In the final analysis, you'd have to say that Fodor won the battle that day, but the connectionists may have gained an advantage in the war. Fodor's talk had a larger meaning for the connectionists: The old guard and the new guard are two very different models of cognition, struggling to win the hearts and minds of scientists. To the new guard, Fodor's appearance at UCSD, like a robin heralding the coming of spring, meant that the old guard is beginning to regard connectionism as a serious threat.

The Language of Thought

Though threatened, the old guard's formalist approach to language and cognition stands on very high ground. Hammered out through decades of research in artificial intelligence and linguistics, it owes much of its theoretical backbone to one man—Noam Chomsky. When Chomsky came upon the scene as a young man of twenty-nine with his *Logical Structure of Linguistic Theory,* linguistics was in disarray.

Linguistics' goal is simple enough to state: determining how people understand one another when they talk. In other words, how are we able to generate sentences that are consistent with all the other sentences that we—and the person listening to us—generate? While it's true that many people don't always speak grammatically, no one ever makes a big mistake such as "Dylan named baby a had Pam" when he means "Pam had a baby named Dylan." Clearly there is *some* kind of regularity and structure there. What is it?

In the early days of linguistics, two general models of those regularities dominated scientific discussion. In one such model, *finite-state* grammars, sentences are formed by picking the first word, working from that to choose a second word, then using that to generate a third, and so on through the sentence. The other model, *phrase-structure* grammars, was based on the observation that in most sentences, various kinds of words seem to be regularly grouped with other words. These groups of words, or *phrases,* can be further broken down into subphrases. Linguists set out to characterize the number and types of phrases in a language, hoping to find a complete set of phrases that could be assembled like building blocks to make sentences.

Chomsky showed that both approaches were flawed. In the case of finite-state grammars, they failed to account for many sentences such as "The book that you ordered as a present for your children this Christmas is arriving today." The subject, *book,* is many words away from *is arriving,* so a finite-state model could not predict whether the verb would be plural or singular. Secondly, Chomsky argued that phrase-structured grammars were also limited in the kinds of sentences they could explain; reworking the grammars by adding more phrases and subphrases would lead to a model so complex as to be useless. While Chomsky couldn't prove this assertion, many linguists were persuaded by his arguments.

Chomsky then offered his own approach to analyzing language. One of the fundamental principles of his model, which he called *transformational grammar,* was the idea that the *syntax* of language—the way the words are put together—could be studied independently of *semantics*—what the words and sentences mean. For Chomsky and his followers, the study of language is a formal study; language is regarded as a system of symbols and the rules that govern how those symbols are manipulated. In that regard, Chomsky's approach to understanding language paralleled the research programs emerging at the same time in other branches of mind studies. As Gardner points out, "Chomsky was a child of the new era of Wiener, von Neumann, Turing, and Shannon."

Chomsky's work forged a revolution in linguistics, changing the way scientists describe and study language, turning linguistics into a science. But while the formal tools Chomsky developed advanced the study of language in a general theoretical sense, researchers who study how people actually speak have found that Chomsky's formal analysis of grammar is difficult to apply to actual human language. Traditional linguists try to explain this difference by citing a distinction between *performance* and *competence.* The job of a traditional linguist, they say, is to describe the competence of the language; that people can't perform that theoretical system very well is due to practical limitations of the brain's memory abilities.

Theoretical Language versus Human Language

To connectionists, however, this lack of agreement between competence and performance is good evidence that the traditional analysis of language is fundamentally flawed. "The proper analysis of language has to involve the language that people actually speak," says Don Norman. "You can't have this performance-competence distinction, because language isn't some abstract thing. Language evolved through the brain, and so our understanding of language has to be built around the kind of mechanism the brain is."

As Patricia Churchland indicated at Fodor's talk, the formal model of how we speak is not without its own problems. For example, according to this model, language is *recursive*—that is, any sentence can be extended to the right, left, or center. A sentence such as "Eric works for a magazine," for instance, can be extended to "Bill thinks that Eric works for a magazine," to "John said that Bill thinks that Eric works for a magazine," and so on, indefinitely. The same is true for extending a sentence to the right: "Eric works for a magazine and has two children and a wife named Ellen and plays football," and so on. People seem to be able to handle these types of sentences to some extent.

But in fact, a sentence can go only so far with left-and-right recursion before people become confused. Furthermore, a sentence with center recursion—"The man the boy the dog bit yelled hit"— can't be understood at all, even though such a sentence is grammatically correct according to the traditional linguistic models. "Fodor tries to weasel out of the center-recursion problem by saying it is just a performance-competence distinction," says Norman. "But I would say that is strong evidence for what *I* believe—that we do not speak the way traditionalists say we do, and that this elegant, high-principled analysis of language is fundamentally wrong. Now, you will find that this raises more emotional overtones than anything you might hear. They will say, 'Old Norman is one of *those* people, one of those hackers and experimentalists or whatever.' "

This distinction between the formalism of theory and the empiricism of experiments distinguishes what is commonly known as the "East Coast" approach to the study of the mind from the "West Coast" approach. Chomsky and other researchers at MIT are at the center of the East Coast view. In psychology, the center of the West Coast view is UCSD. But when it comes to language, the center is Berkeley.

The West Coast View

The University of California at Berkeley seems to attract people who make careers out of challenging traditional views of the world. So it was no surprise that when Terry Sejnowski gave a guest lecture on connectionism there, the room was packed with listeners. In fact, the interest was so great that Sejnowski gave two lectures at the university—one to a group of engineers, the other to linguists.

The fact that Sejnowski can talk to both groups demonstrates his ability in mental hopscotch as he roams new territories, spreading the word on neural networks and searching for applications and ideas. But it is also a demonstration of connectionism's widespread impact on the scientific community at large, one that is both exciting and dangerous.

The history of science is not one of gradual, steady progress, but of periodic bursts of creativity and new understanding. The computer revolution in the 1950s brought fresh new ideas to all fields of mind research. Now many researchers, frustrated by a lack of progress in artificial intelligence, are looking to connectionism for new inspiration. But while there may be many reasons for rejecting the traditional model of the mind, reasons for embracing connectionism are still mostly hypothetical. Connectionism *looks* promising, but it comes in many flavors these days, with a plethora of designs, philosophies, and styles. Much more research is needed to decide which theories and models hold the most promise. In the meantime, there is a risk that misunderstandings about one aspect of a particular model will lead to misunderstandings about connectionism in general.

When Fodor talked about neural networks in his lecture at UCSD, for instance, he used a "single-node" model of a network, similar to those developed by Jerry Feldman. Fodor assumed that the model was representative of neural nets as a whole, which is far from being the case. "The little diagrams Fodor drew have little to do with connectionism," says Sejnowski. "The distributed representations that I use in my models are very different from the networks Feldman uses."

Connectionism's departure from traditional modes of scientific inquiry also makes it difficult for researchers outside the field to analyze neural network models. In traditional cognitive science, a *theory* of how the mind works is distinct from the *machine* the theory is tested on. The details of the computer hardware that implements the theory are considered unimportant because at a fundamental level, all computers are alike. But in neural networks, the role of the machine is *not* trivial. In a neural net, theory and practice are intimately intermeshed. The theory is represented by the structure of the network itself and is therefore limited by the details of the machine's architecture.

This lack of distinction between theory and practice changes the rules by which scientists assess the worth of a particular neural net project. For example, two linguists recently released a paper severely criticizing the neural network, designed by McClelland and Rumelhart, that could change verbs into the past tense. The linguists had a long list of complaints: The model is incapable of representing certain kinds of words; it doesn't explain patterns of psychological similarity among words; it models many kinds of rules that are not found in human language, doesn't capture generalizations about English sound patterns, can't model general processes of word formation, doesn't work on certain words. In short, the model doesn't address vast areas of human language abilities. On the basis of these criticisms and others, the linguists concluded that the idea that neural networks can help scientists understand language is "unwarranted or at best grossly premature."

But McClelland and Rumelhart's model was, in a sense, a "Model T" version of neural networks. Built before the back propagation learning procedure was developed, it had no hidden units and only a small number of neurons. Despite these limitations, however, the network proved to be very adept at making generalizations about a particular (albeit quite limited) aspect of language. It was not intended to be a complete model of language acquisition in children, but to demonstrate that much of what linguists had long assumed to be the result of children learning rules for processing language might actually be the result of neural networklike computation.

"If there was a problem," says Rumelhart, "it was that the network was *too* good. It was as good as or better than the other models at accounting for kids' errors. So people took every little detail of the model very seriously, rather than seeing it as illustrative of a new paradigm for accounting for these things. Most of the linguists' paper is a diatribe against bits of the model that are unimportant to the model's fundamental behavior, or asking things of the model that are obviously out of its scope."

Part of the reaction, says Rumelhart, is due to many linguists working in the formal tradition of language analysis. "They don't have a clear understanding of model building, as opposed to writing general theories. The tradition in linguistics is to generate theories where a single counter-example is enough to knock down a whole theory. So with our network, they miss the point. They say that if you don't believe this model in every detail, then there's no reason to pay attention to anything that connectionists do. It's as if the whole question of whether connectionism will contribute anything to understanding language depends on how good this one neural network is."

The backlash generated by McClelland and Rumelhart's model of verb learning is a demonstration that there are many other researchers who, like Fodor, find the sudden interest in connectionism irritating, perhaps even threatening. But Sejnowski is adamant about downplaying the notion that connectionism will soon be overthrowing the traditional model of the mind. "Such a change may come someday," he says, "but right now we don't even know what to replace the traditional stuff with, so I wouldn't go around saying that connectionism will overthrow this or that. That's too simplistic, and all it will do is get people very defensive. A better way to look at it is that we are adding a new dimension to the study of the brain and mind." But among linguists raised in the tradition of Chomsky and Fodor, the general rule is that you don't walk away from a fight. "For connectionism," says McClelland, "language is the place where the heat is the hottest."

One of the reasons for the heat in linguistics is that the traditional approach has been under fire ever since Chomsky brought out his first theories. Not only have linguists found that Chomsky's formalism doesn't apply very well to real situations but other linguists have long been skeptical of Chomsky's claim that the syntax of language is isolated from the meaning of words. For example, the formal analysis of language doesn't explain how we are able to understand the differences between sentences such as "Tom threw the ball for charity" and "Tom threw the ball for the baseball scouts," or "Trudy saw a cowboy with a horse" and "Trudy saw a cowboy with a telescope."

The Iconoclast Linguist

One linguist who has been searching for alternative ways to explain language is Berkeley's George Lakoff, who was in the audience as Sejnowski gave his talk. Lakoff had been an increasingly frequent visitor at connectionist gatherings, cornering people like Jay McClelland during a lunch break at the Cognitive Science Conference, meeting informally with Dave Rumelhart, and helping Patricia Churchland conduct a symposium at UCSD.

Lakoff is a genial man who enjoys dining at Chez Pannisse, the Berkeley restaurant famed for its California nouvelle cuisine, almost as much as he likes poking holes in Chomsky's theory of grammar. "You have to remember, I was in the first generation of Chomsky's students," he says, sampling the pear sorbet after lunch at his favorite restaurant. "I was one of the people who did the early work in those years, one of the people who got the first batch of results that supported Chomsky's transformational grammar."

But Lakoff soon began to suspect that Chomsky's theory that the meanings of words are unrelated to the syntactical structure of language was incorrect. "Meaning *does* influence syntax," he says. "Of course, once I realized that, I was excommunicated."

One reason for Lakoff's rejecting the formalist approach to language arises from his own work on how we make categories. According to Lakoff, putting things in categories is a major part of our thinking process. We put things in a category when we recognize things—realizing, for example, that the approaching animal that looks a little like a lion, but has striped fur and no mane, is probably something we want to avoid anyway. We also put things in categories when we think of processes; a rock can be thought of as belonging in the category of "things good for hammering," not because it looks like a hammer, but because it shares some qualities with hammers, such as hardness.

In Lakoff's *Women, Fire, and Dangerous Things,* whose title comes from an Australian aborigine category that contains all those items. He argues that traditional ideas of categories, as conceived by many cognitive researchers and philosophers, don't apply in the real world. Traditional philosophy, what Lakoff calls *objectivist* philosophy, assumes that the world is made of objects that have certain properties and relationships with other objects. Tables, chairs, and coffee cups, for example, are objects with certain properties; the coffee cup is white, the table and chair made of wood. These objects are also related to one another in certain ways; the coffee cup is on the table; the table is near the chair.

Furthermore, this objectivist philosophy assumes that the world can be described as a collection of objects, their properties, and their relationships. A traditional linguist uses this philosophical view in analyzing language. A sentence is a series of strings of *uninterpreted* symbols, and those symbols are given meaning by defining the words *coffee cup,* for example, as "the white object on the table with coffee in it." "To a traditional linguist, the real world looks just like a computer database," says Lakoff. "Seventy-five percent of the people in cognitive science today think that this is how it works."

But Lakoff's research challenges this view that meaning is based on an objective truth independent of human minds. Rather, Lakoff believes that the meaning of words relates to the nature of ⅛ the human organism—the fact that we all have a body with a brain.

The traditional model defines categories as a set of things that share the same properties. The boundaries of categories are sharp; if an object has the required set of properties, it is in the category; if it doesn't, it's not in the category. Because all the members of a category share certain attributes, no particular member is more representative of a category than another. One pencil, for example, should be equally as representative of the category *pencils* as another. Since symbols have no meaning of their own, the symbols we use to represent categories are arbitrarily assigned; nothing about the world or our mind and body determines what those categories are or what goes into them.

But this classical view of categories breaks down in the real world. Even simple objects make the task of listing the properties of a category a nightmare. For example, it's difficult to list the properties that make a chair a chair. We all seem to recognize one when we see it, but a physical description—"It's got four legs"—or a description of its purpose—"We sit on it"—doesn't seem to be enough to contain all the things we call chairs while excluding the things we don't call chairs. The boundaries of categories also seem to have fuzzy edges, with one type of object gradually crossing into another type—something experienced by anyone who has put a cigarette out in a candy dish.

Clues from Color Perception

Further evidence against the traditional model of categories comes from the pioneering work of psychologist Eleanor Rosch. Rosch tested the color-naming abilities of a tribe in New Guinea that uses only two names for colors—*mola* for bright, warm colors; *mili* for dark, cold ones. Showing her subjects a variety of color samples, Rosch asked them to look at one color, wait half a minute, and then try to match their memory of that sample to one from a batch of forty colors. She found that the subjects could consistently remember certain colors better than others. Also, they made the same kinds of correct answers and mistakes that a group of English-speaking people made on a similar test. In other words, though the New Guinea people had only two names for colors, their ability to *remember* colors was similar to that of English-speaking people who have many names for colors in their language.

Rosch's work was a challenge to classical notions of categories because according to the classical view, the divisions of the light spectrum into different colors was arbitrarily decided by a culture and reflected in the color names of its language. But Rosch's work showed that the way people made categories of colors in their *mind* was independent of how they named colors.

Her work was given further credence by the research of anthropologists Brent Berlin and Paul Kay. It had always been assumed that because categorizing was an arbitrary process, different cultures would naturally divide the color spectrum in different ways. Kay and Berlin took 320 swatches of color, like those found in paint stores, and traveled around the world asking people to point out the "best blue" or the "best red" in the samples. In every culture, people selected nearly the same swatches as the "best" example of a particular color. Even in cultures that had only one word for a broad group of colors—*grue,* for example, for the color spectrum ranging from green to blue—the "best" example of grue was the green or blue that people in other cultures had picked as "best." Kay and Berlin also found that there seemed to be a worldwide structure in the way colors were named. In cultures with only two names for colors, for example, the names stood for "dark" or "light." If there were three color names, the third name was for "red." If there were four color names, the fourth name was "yellow" and the fifth was "green," or vice versa.

These anthropologists' work suggests that rather than arbitrarily dividing up the color spectrum, we categorize colors according to some universal structure. That structure, says Lakoff, may well depend on how our brains are constructed. In our visual system, some neurons deal with red and green; others, blue and yellow; and others, dark and light. When you look at the "best" example of blue, suggests Lakoff, the neurons fire in a particular optimum pattern common to the brains of everyone. Other "best" examples of color have other specific patterns of firing. "The color red isn't a property of the outside world," says Lakoff. "It's a property of the mind."

Rosch and other researchers have found that there are many categories that seem to have "best" examples. For instance, when most of us think of birds, we think of something that looks like a robin. Imagine the difference in your reading of the sentence "The birds perched on the windowsill" if the word *birds* were replaced with the word *turkeys*. Likewise, a four-door sedan somehow seems more representative of *cars* than a Land Rover, and Hugh Hefner typifies bachelorhood a little better than the pope.

These "best examples" are called *prototypes;* many cognitive researchers believe that instead of making categories by listing an object's properties, we often use a prototype to define a category, then assess whether other items belong in that category by comparing them to that prototype. We use a prototype dog, for example, to categorize all other four-legged creatures. "I got into a big discussion with a colleague of mine," says Jim Anderson, "about whether a dog with three legs was really a dog. He was arguing that it wasn't, which was pretty amusing, because at the time there was a dog with three legs roaming around the campus."

Anderson's work suggests that neural networks may also use prototypes to define categories. In one experiment, he took a pattern of dots and made a set of new patterns by randomly distorting the arrangement of dots in the original. This new group of patterns formed a "category" of patterns, which he trained his neural network to recognize. Anderson didn't include the original dot pattern in the examples he used for training the network. But when he retested the network's ability to classify the patterns into the proper category, he found that the pattern most rapidly categorized was the *original prototype*, though the network had never seen it before. When Anderson reproduced his experiments with human subjects, he got similar results.

Not only do some categories have prototype members that are more representative than others but categories themselves appear to be arranged in a hierarchy where some are more representative than others. For example, Berlin examined the ways that a dozen primitive societies named the plants around them. He found that they didn't organize the flora according to a particular use a plant had in their culture, such as whether it was edible, dangerous, or useful for shelter; instead, the native people named the plants in roughly the same way that Western science categorizes them, based on similarities in appearance. Furthermore, even though these cultures had many different levels of names for plants, usually they used the name of the plant that corresponded to the category closely matching the botanical level of *genus*. For plants, at least, there seemed to be a basic level of categorization that all people are most comfortable with. "We seem to operate at a 'natural' level of complexity," says Anderson. "We say in normal speech, 'Look at that robin on the grass,' as opposed to 'Look at that organism on the flat area of Kentucky bluegrass, clover, and creeping red fescue.' "

Fundamental Categories

According to Lakoff, we have such basic levels of categorization because the brain is part of the body. If categories are independent of the brain, there is no reason why we should form these basic-level concepts such as *chair* (as opposed to *furniture*) or *car* (as opposed to *vehicle*). Yet we do have these basic-level concepts. If asked about a typical *car*, for example, you can form a mental image of one and describe it in great detail. But it is much harder to give a detailed description of a *vehicle*, because the category is so broad that there is no mental image that includes cars, ships, and planes.

The reason we use cars as a basic category, says Lakoff, is that basic-level categories are centered around human use (you steer cars and you sit in chairs), and that interaction determines how you organize your knowledge of the world. Children begin to organize

information at this basic level, then generalize upward and special-
ize downward.

While Lakoff has dedicated most of his career to proving the
classical model of linguistics wrong, he has only recently begun to
turn to connectionism as a possible way to construct an alternative.
"I'm considered a renegade—a well-known renegade," he says.
"My work has taken me indirectly to connectionism." Part of his
motivation to explore neural networks is what motivated other
researchers in many other fields: he became frustrated with trying
to use traditional theoretical methods of analysis to solve longstanding
problems that arose in the real world. "Linguists got into linguistics to study not theory but *languages*," he says. "They are studying Burmese, African languages, and American Indian languages.
They are up to their ears in the real details, and they *know* that this
Chomsky stuff doesn't work. Researchers want to work with the
data they've found, and they are looking for an alternative model.
That's something most people don't realize—there is a dam about to
break, and there is going to be an incredible backlash when it does."

Lakoff concedes that neural networkers still have many problems to overcome before they will be useful to linguists. For
example, they must be able to represent a concept like "the blue
tablecloth," linking the concept "blue" to the concept "tablecloth."
That's difficult to do in a distributed neural net where large groups
of neurons are responsible for holding information. Since the memories in a neural net are represented by a *pattern* of activity among
the neurons, connectionists have to find a way to incorporate an
activity pattern for "blue" into an activity pattern for "tablecloth."
As Fodor points out, connectionists are also struggling to make their
machines do simple logic, to represent relationships like "John
loves Mary," and to handle recursive sentences like "I heard that
Paul said that Heddy knew about John's marriage to Sarah."

But these complications are not enough to turn Lakoff away
from connectionism. "Is there a theoretical barrier to these problems?" he says. "I don't think so. But that is just a guess. I can't
rationally give you a defense. These are well-known problems that
have been already solved by classical symbol-manipulating systems. But the classical systems have trouble for *a lot* of other
reasons. As far as I'm concerned, the classical systems can't do
language at all. Connectionists should still go after language, because it is the only hope we have."

Language and Connectionism

For Lakoff, connectionism's appeal is that it may be able to connect the language produced by our minds with the structure of our brains. "What's intriguing is that while in traditional AI there is only *description,* connectionism might give us *explanation.* Connectionism's power comes from its potential to explain *why* language looks the way it does. And that is really exciting."

For example, given a connectionist framework of the mind, it makes sense that there would be basic-level categories and concepts. As the brain juggles inputs from neurons responsible for perception, motor control, and other areas, various patterns of neural activity interact with one another. The overlapping activity may be what leads to the formation of a general pattern representing a basic-level concept.

Connectionism has the potential to explain other characteristics of language as well. The language-of-thought model of cognition doesn't include the idea that we learn concepts throughout life; it is assumed that abstract concepts are innate and that in different cultures, some concepts are activated and others are not. This leads to the somewhat bizarre conclusion that we are born with concepts such as "latitude" and "second baseman."

Connectionism may provide a way to explain how we learn new concepts. Consider the abstract concept "love." In the traditional model, our concept of love is innate. Lakoff's work, however, suggests that we learn to understand a complex concept such as love by using metaphor; that is, by linking it to another, more familiar concept. One such metaphor is the journey. When talking about a love relationship, says Lakoff, people often use expressions such as "we can't turn back now," "we're spinning our wheels," "we're off the track," or "on the rocks." In these expressions, the relationship between two lovers is thought of as a vehicle, and the lovers are thought of as travelers. Their path is the path of common purposes, and along that path are impediments to the journey: crossroads, wrong turns, and roadblocks.

According to Lakoff, we understand one concept—love—by mapping it onto another concept that we experience in the world—a journey. Connectionists may have the apparatus to explain how that mapping occurs, says Lakoff, because it suggests that patterns of activity that fit one concept may be overlaid to a new concept.

Connectionism also suggests that it is easier to learn a variation on an old concept than to learn a totally new one. New words sometimes appear in a language, but often new concepts are named by modifying an old word. Chomsky's phrase "transformational grammar," for example, is a variation of the old word *transformation*. "Chomsky could have made up a totally new word like *ziglot*," says Lakoff. "But he took a word that was already there—a word that had a meaning systematically related to what he was talking about—and created a new sense of it. That's the normal way these things happen."

Consider the word *took*. You can say, "I took the pen" or "I took the pen to you," as well as "I took Mary to the movies," "I took a punch," "I took it all in," or "I took a look." All these uses of *took* have slightly different but related meanings. Piggybacking a new meaning on an old word isn't explained by the classical approach to language, but connectionism suggests that this is how language *should* change; when a neural net is trained on new examples, it adds new connections over a pattern of neural connections that already exists. "The classical approach views the lexicon as simply a big list of words," says Lakoff. "It doesn't explain why words should be related."

In the classical model of the mind, things not predictable by rules are considered arbitrary. That view derives from the classical definition of *reasoning:* deducing one thing from other things. But connectionism suggests that reasoning is something else, says Lakoff. "In connectionism, reasoning is not deducing one thing from another, but rather *putting things together well.* That changes the whole idea of what it is to think. And that means philosophy will change, economics will change, sociology will change, and anthropology will change."

One problem in anthropology, for example, is the question of how conceptual systems arise in different societies. If it turns out that the process is not an arbitrary assigning of concepts to the world, that people use a particular conceptual system because they have a brain of a certain kind, then scientists will have to go back and ask those anthropological questions again. "The social sciences are beginning to die out in a lot of ways," says Lakoff. "Connectionism may be a way to reconstitute them. Of course, humans are much more complicated than neurons. But connectionism may provide a way of modeling how people represent themselves and their cultures."

The Culture Connection

A researcher who is trying to see how connectionism might provide insights into the origins of culture is Ed Hutchins of UCSD. Hutchins looks a little like a surfer. His thick blond hair, swept back, nearly reaches his wide shoulders, and his face wears the tan of someone who has seen a lot of sun.

That description would probably exasperate Hutchins, not because the physical details aren't accurate but because the conclusion says more about the culture of the person who is describing him (midwesterner, landlocked) than the person being described. Hutchins is more sensitive than most about that kind of bias because he is an anthropologist. Anthropologists try to make sense of other people and their cultures, though they often find themselves tripping over their own cultures in the process. Hutchins found evidence of anthropological biases when he went to Papua New Guinea to study transactions in land rights. "There was a longstanding claim that primitive technology meant primitive thought, that these people couldn't reason," he says. "But nobody had ever thought to study how those people reason about things that they *cared* about."

Usually, such studies are conducted by an anthropologist who designs an experimental task and asks subjects to solve it. This puts the burden on the experimental subjects to determine the overall structure of the task. Hutchins took a different approach and let the tribes in New Guinea design a task, putting the burden on himself to discover what they were doing. Studying the New Guinea transactions and disputes over land rights, Hutchins found that the natives used the same kinds of reasoning found in a Western society.

Hutchins is not an ordinary anthropologist, and so he fits right in with the PDP group at UCSD, where he was a graduate student and is now an assistant professor. "I'm not really in the mainstream of anthropology," he says. "I'm pretty fringy." Hutchins is a *cognitive* anthropologist. While most anthropologists try to describe the knowledge an individual needs to function in a culture, Hutchins explores the computational problems people face, the representational structures they use to solve these problems, and the tools and resources they employ to get the task done.

One of his studies examined how sailors in Micronesia navigate across the open ocean, without instruments, to reach a tiny coral island some five hundred miles away. Hutchins found that the navigators shifted their frame of reference in a way that might seem bizarre to a westerner. Instead of thinking of themselves as traveling across the ocean, the navigators imagined that they were stationary and that islands over the horizon were moving past them.

Hutchins's paper on the Micronesian navigators, entitled "Why the Islands Move," was coauthored by Geoffrey Hinton. Hinton first interested Hutchins in connectionism while they were graduate students. Hutchins saw parallels between his interests and what Rumelhart was trying to do with his schema ideas of how people understand stories. He attended a seminar taught by Rumelhart and became interested in artificial intelligence and cognitive science. Then he met Hinton, who was working on associative memory models, and attended the connectionist meeting organized by Hinton and Anderson. He has been interested in connectionism ever since.

The connectionist modeling of the brain and mind might seem a far cry from the study of cultures. But in the same way that connectionism may bridge neuroscience and psychology, it may provide a link between the study of individual minds and the study of groups of minds and their culture. If a group of simple neurons can work together to produce an intelligent brain, individuals with a limited ability to do a task might be able to do it better as a group. "I'm interested in systems composed of many people," says Hutchins. "If I had a wish list—if I could say, 'Gee, I wish that AI would have a paradigm that looks like such and such'—then connectionism comes close to what I would wish for."

In many cultures, the performance of a cognitive task such as navigation or engineering is distributed among a group of people. Even someone acting in isolation is usually working with technological partners; most journalists, for example, use a pencil, notebook, and often a tape recorder. These technologies influence the structure of the journalist's cognitive tasks during an interview. A journalist might take different kinds of notes if a tape recorder were on at the same time, and if the journalist had neither a tape recorder nor a notebook, the cognitive work would be very different.

In fact, a single brain can increase its power many times over by using technology. Even a few sheets of paper and a pencil will give a brain much more memory and allow it to do mathematical problems serially, something it finds very difficult to do on its own. A technology like writing can go further to change the structure of computation. When you take a book from a bookshelf, you are communicating with a person who may have died hundreds or even thousands of years ago. With the influx of electronic media, the computational structures of cultures change even more.

Being aware of the environment in which cognitive tasks are performed gives Hutchins a different perspective on neural net modeling. "Characterizing the environment in which neural nets learn is an important thing that many people in the connectionist community usually overlook," he says. "They just blithely write a program that produces the environment the network learns in, without ever asking where the *structures* in that environment come from in the first place."

Because neural nets can model environments, connectionism may help anthropologists discover the origins of social structures. In conventional artificial intelligence, it's difficult to represent how people learn to adapt to their environment or how a computational system, like a set of navigators, interacts with its culture. But these kinds of processes flow naturally from connectionist models. "As soon as you start talking about connecting entire neural networks to each other," says Hutchins, "it begins to have a real social smell to it."

In fact, answering questions about how a group of brains interact might be easier than trying to discover how a group of neurons behave in the microscopic world of brain tissue. "I have an advantage," says Hutchins. "A typical connectionist can't step in between the ears and say, 'What is the language of representation? What's the structure of these representations, and how are they transformed?' But for me, a group of navigators is my cognitive unit. I can step into the middle of this group and look at the representations they are passing to each other. I can step into it in a way that a psychologist would just love to be able to do with a brain."

Anthropologists have long been reluctant to attempt to explain the origins of features they find in a culture. Most anthropologists simply try to describe the regularities that emerge in a culture and understand them in terms of simple laws. "There's a real antireductionist firewall built into anthropology," says Hutchins. "It's probably there for a good reason, too, because we *don't* have the tools that allow us to see how it might work." But connectionism may provide anthropologists with these tools because one of the most important principles behind neural networks is the ability to generalize. As a neural net learns to perform a task, it makes generalizations about the examples it is trained on—how a group of similar verbs change from past to present tense, for instance. These generalizations reveal the underlying structure in a neural network's "environment" of examples, a common thread that connects various elements in a training set.

In a similar way, anthropologists may be able to use connectionism to explore the emergence of structure in cultures. A society might be seen as a system that goes through an adaptive, "constraint satisfaction" when it processes ideas in its culture. In a sense, people in a culture constitute a special kind of "medium" for ideas to grow and evolve in. In this medium, an idea competes with some ideas and reinforces others, resulting in a complex interactive system that eventually settles into a stable state of interacting ideas that defines a culture.

How a society of minds joins together to produce such a culture is being studied by Aaron Cicourel, a colleague of Hutchins and a sociologist at UCSD. Cicourel studies the kinds of reasoning physicians use to make diagnoses, research similar to Hutchins's study of navigators. As in anthropology, the traditional approach to sociology has been to describe regularities on a social level, not how these social structures might have arisen. But connectionism may help sociologists because, like a neuron in a network, each individual in a society is constrained by the actions of others. A child taking an exam, for example, brings not only a set of knowledge to the test but also a vast web of social connections that may subtly influence the outcome.

Besides being somewhat on the fringes of their own fields, Lakoff, Hutchins, and Cicourel are outsiders among connectionists.

The "new" connectionist movement arose through the efforts of psychologists, neuroscientists, and engineers, so the focus of most neural net research has been on the mechanics of the brain and mind rather than language, culture, and society. But no brain operates in isolation. Understanding the activity of groups of brains—how they communicate with each other, how they join to solve problems, and how their interactive behavior produces culture—may be as important to understanding ourselves as research on the actions of neurons. "There are some people who think that we should really understand origins of the mind and brain before we start worrying about understanding *groups* of minds and brains," says Cicourel. "But how can you find out where the hell you've come from without trying to discover where the hell you've *come?*"

Such a broad-reaching interdisciplinary program will take shape as more and more researchers from various fields become acquainted with the new model of the mind. A connectionist intent on spreading the word is Terry Sejnowski, whose training in neurobiology, physics, and computer science enables him to traverse a wide spectrum of scientific disciplines. Sejnowski created a simple but brilliant demonstration of the enormous potential of neural networks, a model that a scientist in any field could grasp. His machine, "NETalk," learned to read aloud.

CHAPTER NINE

THE PRIZE PUPIL

NETalk Learns to Read Aloud

Q: How many connectionists does it take to screw in a light bulb?
A: If you wire it right, you don't need a light bulb.
—Overheard in a discussion among graduate
students at a cognitive science conference

Creating a neural net that can read aloud might not seem like much of an achievement in a world where computers control fighter jets, help perform surgery, and find deposits of ore. Terry Sejnowski's interests, however, go far beyond making machines that talk. He is interested in how the brain works, how the billions of interactions among neurons produce the remarkable abilities of the mind.

To discover the mechanism of the mind, says Sejnowski, researchers have to take into consideration that cognition is done by a living brain. "Many cognitive psychologists have bought— hook, line, and sinker—this assumption that you can analyze thinking at a conceptual level independent of the machines that are implementing it," says Sejnowski. "That's the fundamental assumption in artificial intelligence. And there's a certain range of problems where you can get away with it. But how far can you go? I don't want to know what a machine is capable of doing if it had all the time in the universe. There's a *biological* constraint. A machine should be able to do the computation in a time that is comparable to human time. If that's your constraint, then the architecture of the machine makes all the difference in the world."

Of all the architectures for thinking machines, the most prevalent is the nerve tissue in your head. "Since I'm interested in studying the human brain," says Sejnowski, "I have to confine myself to designs that are brainlike, machines that work with the same assumptions that biology uses when it is coming up with a real brain. Not only do I have to solve the problem, but I have to solve it while working within the same confines that nature had, in terms of components and time."

Sejnowski seems always to be in a hurry. He fidgets, jiggles his knees, and nods with a quick "uh-huh" while you talk. He often grabs a pen out of his pocket and waves it around as he speaks; sometimes he unknowingly waves two pens at the same time, one in each hand like a chain smoker. Sejnowski is probably better described as a chain thinker. He wants to move on, get to the next problem, explore the next idea. As he demonstrates NETalk's speaking abilities, a proud smile keeps welling up and spilling over his necktie. "NETalk is what I call an *exemplar*, a demonstration of what neural nets can do," he says. "It's not meant to be a detailed model of exactly what happens in your brain. After all, the network has only three hundred neurons."

Sejnowski started his academic career as a theoretical physicist at Princeton University, studying the physics of black holes under the tutelage of the famed cosmologist John Wheeler. But while immersed in the cold mathematics of physics, he became drawn to the warmer realm of the living brain. "To a theoretical physicist, the brain is very appealing," he says. "You can't see a quark, you can't do an experiment on a quasar. But in the case of the brain, you can actually hold it in your hand. You can poke electrodes in it and ask questions about it. It's a mystery, but it's accessible."

Sejnowski finished graduate school with a Ph.D. in physics, but by then, he had strayed far from relativity. His dissertation, sponsored by John Hopfield, then a professor at Princeton, explored some of the physical properties of neural nets. After completing his thesis, Sejnowski was completely won over to the study of the brain. He eschewed the traditional postdoctoral routes in physics and instead spent a year in Princeton's biology department and three more years studying neurobiology at Harvard University. There he realized that thinking about thinking requires getting your hands a little dirty. "It's pretty much hopeless to guess how the brain works based on pure thought," he says. "Nobody is going to sit down in a room with a paper and pencil and understand what is going on, for the simple reason that biology is not always elegant."

Scientific elegance is rare in the land of the living because evolution is not a process whereby creatures develop toward perfection. Rather, evolution occurs randomly and erratically as individuals take advantage of genetic mutations that make them better adapted to their environment. "A lot of the details and organizational decisions in biology are historical accidents," says Sejnowski. "You can't just assume that nature took the simplest and most direct route to do something. Some features are remnants of some earlier stage of evolution, or it may be that some genes that happen to be around are commandeered for some other purpose. So to figure out what's going on in the brain, nothing replaces going in and looking at it."

During his apprenticeship as a neurobiologist, Sejnowski did just that, exploring the brain's neurons, chemicals, and structure. But he found that this "wet" approach has its own problems. "I discovered that it was *also* hopeless," he says, "to think that purely on the basis of accumulating more and more facts about the brain, you would be able to build up an understanding of how it is organized as an information-processing device." Because the brain is a living organ that does more than process information, neuroscientists risk getting bogged down examining extraneous details concerning the metabolism of nerve cells rather than trying to understand how those neurons perform cognitive tasks. Only a small fraction of the total neural activity in the brain may be crucial to the information processing involved when you recognize your mother. The only way a scientist can distinguish that small percentage of relevant information from the rest of the brain's activity, says Sejnowski, is to have an idea about the organization of the brain as a whole.

Studying the Brain from Both Sides

Sejnowski went through two phases in his career. First he tried to learn about how the brain works on a theoretical level, using the top-down approach to break large problems into smaller pieces. "It's necessary to have some sort of theoretical model to refine your thinking," says Sejnowski. "That way, when you go back to the brain's actual neurons, you know what to look for." Ultimately, Sejnowski found this approach was inadequate, so he tried the experimental bottom-up approach of trying to unite the separate components of the brain into a whole structure. But he decided that was hopeless, too. "Now I'm trying to synthesize a new perspective," he says. "I'm trying to do it from the middle out."

To do that, Sejnowski borrows from the best of both worlds. Analyzing our ability to read a sentence aloud, for example, the top-down approach can be used to examine the features of our brain's output; that is, the sounds we make when we are talking. In the same way that physicists have broken down substances into atoms, and atoms into protons, and protons into quarks, linguists analyze and catalog the components and subcomponents of the sounds of spoken words. They've found that there are some fifty distinct speech sounds used in English and other languages; these are called *phonemes*. There is a phoneme for the "sh" sound; another for a hard "k." Put a string of phonemes together, and you have the sound of spoken language.

The top-down approach can also be used to analyze the input for reading aloud—how sentences are made of words, words of letters. At its basic level, written English is a cipher of symbols; the pronunciation of a word is encoded in the letters and words themselves. Our brains read the code and use it to guide us in pronouncing the output—speech.

But while the top-down approach can help analyze the overall features of text and speech, it doesn't really explain the mechanisms in the brain that guide us when we look at a sentence and read it aloud. "The problem of translating written English words into English sounds is a well-defined problem," says Sejnowski. "You can state what the inputs and outputs are very simply: There's no doubt that I'm looking at strings of symbols; there's no doubt that I'm making these sounds. How does one lead to the other? That's our starting point. That's something to solve."

One way to read text aloud, in fact one that commercial text-reading machines use today, is to store a large list of words and their pronunciations in memory. The machine matches a word in the text with a word in its memory to produce the proper pronunciation. If the machine encounters a word that it doesn't have in its memory, it uses a set of pronunciation rules—such as "if you see *sh,* make a 'shhh' sound"—to guide it. The machines work very well, but they don't really sound human. One problem is that we often pronounce a word differently, depending on the words that come before and after it. "If you look at real speech, the first thing you discover is that the dictionary lies," say Sejnowski.

"Nobody pronounces words according to the dictionary, except words spoken in isolation."

Because a word can be spoken several different ways, using a pronunciation dictionary requires a lot of computation. Suppose, for instance, that a word's pronunciation depending on the word that appears before and after it. Then, for a machine to find a word's pronunciation, it must look up not just one word but a *group* of three words. Sifting through a pronouncing dictionary of 200,000 words may not be too time-consuming, but sifting among the 10^{15} possible three-word combinations of those 200,000 words can be. The task gets even more complex when you consider that sometimes the pronunciation of words at the beginning of a sentence can be influenced by a question mark at the end. There's a significant difference between the way we pronounce the words in the statement "I'm to blame for this" and the question "I'm to blame for this?"

To make things even more difficult, consider the first lines of the poem "Jabberwocky," from Lewis Carroll's *Through the Looking Glass:*

'Twas brillig, and the slithy toves
Did gyre and gimble in the wabe:
All mimsy were the borogoves,
And the mome raths outgrabe.

Here, we could not possibly read the words aloud by looking them up in some sort of pronunciation dictionary in our head. The words do not exist in English! Yet most speakers will read the text in the same way. That we all tend to pronounce nonsense words similarly might suggest that we have a set of general rules for pronouncing letters and words. There are many regularities in the way we pronounce written words, and for the last several decades, linguists have tried to discover rules that capture those regularities. But as Lakoff points out, linguists' rules don't cover all the exceptions.

This variability in the way words are pronounced has stymied the efforts of linguists to reduce our reading-aloud abilities to a set of rules. No matter which rules they start with, they always wind up with exceptions, and there is no agreement on a universal system. "Trying to reduce everything to rules has been a straitjacket," says Sejnowski. "It works for a limited number of possible combinations and at least tells you what you're up against in English. But if you want to solve the problem of how humans go from text to speech, you are going to have to come to grips with the variability of speaker to speaker and word to word in the same text."

This difficulty can be resolved by using the top-down and bottom-up approaches together. A top-down approach analyzes the input and output, and bottom-up tells you what lies between— billions of interconnected neurons. The communication channels between these neurons are modifiable, so by changing the strength of these connections, we physically change our minds when we learn something. Each brain carries a unique wiring pattern that reflects individual experiences.

This combined top-down–bottom-up approach gives an overall picture of what's going on in the machinery of the mind when we read aloud. The brain has a place where the input, the text that's being read, comes in and a place where the output, speech, goes out, and it has a lot of interconnected, communicative, and highly adaptable neurons in between. If these adaptable interconnected neurons make our brain so powerful, Sejnowski reasoned, it might also make a brainlike machine powerful.

So NETalk was born.

Inside the Neural Net

NETalk has a "window" through which it looks at seven letters of a text at once, shifting one letter at a time to the right as it reads. NETalk's output consists of the phonemes that linguists use to represent the sound of speech. The phonemes, played through a device that reproduces the sounds through a loudspeaker, give NETalk a voice. Between where the text comes in and the phonemes come out is a layer of interconnected adaptable neurons. These neurons lack the complexity and detail of real neurons, but are similar in their overall behavior. Each receives signals from a number of neurons connected to it and decides whether to pass on a signal of its own.

NETalk reads a text through a set of 203 text-input neurons—29 neurons for each opening in the seven-letter window. NETalk's reading isn't reading in the visual sense; the letters and spaces of the text are typed into the machine, and each of the 29 input neurons is responsible for signaling the presence of a particular letter or punctuation mark in the window. At NETalk's other end, 26 output neurons are responsible for selecting a phoneme.

The core of NETalk is a layer of 80 hidden units that neither read text directly nor produce the phonemes. Each of the 203 text-input neurons is connected to *every one* of the 80 hidden neurons, and each hidden neuron is also connected to each of the 26 output neurons. Thus, between the text "reader" and the layer of hidden neurons there are 16,240 connections, and between the hidden neurons and the output layer there are 2,080 more—a total of 18,320 connections. In addition, each of these 18,320 connections is *weighted;* that is, its connection strength is represented by a number like -1.5, 0, or 2.2. The higher the connection weight, the more important the connection and the more its influence.

Each hidden unit takes all the information coming in from each of the text-input neurons, considers the strength of the connections of each input, and takes on a numerical value of its own. This number is then passed on to every neuron in the output layer. NETalk's output layer of neurons, summing the information from all of the hidden neurons, selects a phoneme. It might seem extremely complex, but that's what you'd expect of a machine that's modeled, even remotely, after the brain tissue in your skull.

This layer of hidden neurons gives NETalk the power to read aloud. "Hidden units allow you to think a little between the inputs and the outputs, if you'll allow me to stretch the word *think* a bit," says Sejnowski. In contrast, imagine an animal with only input and output neurons. The input neurons might be those in its eyes or nose, and the output neurons might be the nerves that control its arms and legs. Such an animal would be capable only of very simple behavior. It could learn to do simple things like avoiding harmful food by connecting the input of the neurons that smell a particular food with the output of the action neurons responsible for moving away. But such an animal would have no hope of learning something complex. "If you want to learn language, simple association is not sufficient," says Sejnowski. "You need to have a *brain*. Having a brain means having a set of neurons between the motor neurons and the sensory input." NETalk's hidden neurons enable the network to do sophisticated computations to solve difficult problems because in these hidden neurons, NETalk holds internal representations of the world, just as our brains hold internal representations of faces or actions.

The hidden units may be what makes NETalk so smart, but they are also responsible for the machine's worst headache: How do you give NETalk's hidden units the internal representations it needs to solve the problem of English pronunciation, so that when NETalk reads the word *ship,* it will select the "sh" phoneme instead of an "s" phoneme? It would be practically impossible for a programmer to try to *guess* the best way to adjust the eighteen thousand or so connections that give NETalk its problem-solving power. It would be like trying to sew two hairbrushes together, bristle by bristle.

But there is another way to do it, the way that other thinking machine, the brain, might do it. Though much of our brain is prewired at birth, the neurons adjust their wiring throughout life as we learn. Despite the best efforts of teachers to remind students that "when two vowels go walking, the first one does the talking," or some such rule, we learn how to pronounce written text not by learning explicit rules but by experience. Children listen and read, and they try and are corrected. Somehow, during that learning period, they are able to pick up all the regularities—and the exceptions—in the pronunciation of English without learning them as rules.

Sejnowski and his coworker Charles Rosenberg, a psychologist from Princeton University, decided to give NETalk its internal representations the same way. Sejnowski was not going to wire it by hand—that was far too difficult—but was going to let the network adjust its connections by itself. To do that, Sejnowski used the back propagation learning rule devised by Rumelhart, Hinton, and Williams. The new learning procedure acts like an ideal second-grade teacher. It compares the output generated by the neural net with what it should be. If the output is incorrect, it doesn't just shout, "*WRONG*"; it goes back through the machine and looks at how the machine's wiring might have led to that output—which connections are weighted the highest and therefore have the most influence over the network's decision. It then changes those weights to produce a better match.

Armed with the new learning procedure, NETalk was ready. "In reading aloud, we know what the input is, and we know what the o utput should be," says Sejnowski. "So we fix the input and output, and let NETalk decide how to connect up its neurons to make a match. The real breakthrough in machines like NETalk is not that they have more neurons; it's that they have new ways of *connecting up* those neurons, new ways of learning."

For its first training session, NETalk was given a thousand-word text of a first-grader's recorded conversation. It was also given the sound of that first-grader speaking the text, transcribed by a linguist into the correct phonemes. The text was fed into NETalk's input, and the correct phonemes to produce the child's speech were given to the learning rule for comparison with NETalk's efforts. The text was run through NETalk again and again, and each time the learning rule adjusted the connections between the neurons to make a better match. NETalk was learning to read aloud by itself.

At the beginning, all the strengths of the connections were set at random. Without any specific knowledge to guide it, NETalk just blathered. As it read the training text over and over, however, NETalk began to pick up features of the way we pronounce English, such as the existence of spaces between words and the regularity of vowels and consonants. After a day of training, NETalk could read the text with about a 95 percent accuracy. More important, NETalk performed almost as well on *any* text, even those containing completely new words. "The exciting thing is that once it's trained, NETalk knows what to do, even when it's given inputs it's never seen before," says Sejnowski. "If NETalk was simply memorizing words and pronunciations, it would have a tough time when it came to new words. But by incorporating the patterns and regularities of English, it is able to generalize and produce the correct pronunciations, even though it's never seen the words."

Hidden Representations

In a way, NETalk still behaves in a rulelike manner. But it doesn't have explicit rules such as "When you see a *sh*, make a 'shhh' sound—except in words such as *crosshatch* and *Dachshund*." Instead, NETalk's "rules" exist as patterns of connections in the machine. "It's not like we're throwing rules away," says Sejnowski. "This is a rule-*following* system, rather than a rule-*based* system. It's incorporating the regularities in the English language, but without our having to put in rules."

As NETalk learned, its hidden units organized themselves to mirror the organization of spoken words, transforming the letters in the text into internal representations of different *combinations* of letters. Those representations were transformed again at the output layer into phonemes. "The individual neurons in this network are extremely simple; they basically just add up numbers," says Sejnowski. "We don't specify any of the internal representations. But we don't need to put them in by hand, because they are *created*. NETalk discovers the representations it needs to solve a problem."

After NETalk was trained, Sejnowski probed its neurons to find out how the representations were organized. He fed all the possible patterns of letter combinations into NETalk's input and looked at the activity of the hidden units. Typically, each time NETalk made a decision about how to sound out a letter, the network responded by activating only about 20 percent of its eighty hidden units; the rest were largely inactive. Different letters activated different small groups of hidden units. Sejnowski discovered that patterns of activity in the hidden units were clustered together. The hidden units that were most active when NETalk was processing vowels, for example, formed a group separate from the hidden units responsible for consonants. Within those vowel and consonant groups were smaller groups. The hidden units that processed the various ways of sounding out an *a*, for example, formed one group, as did the patterns for *e, i, o,* and *u, and even y* when used as a vowel. The hidden units that processed consonants were grouped around similar sounds, such as a hard *c* and *k*.

NETalk is too crude to be a proper representation of what is actually going on in your brain's neurons when you read aloud, but it demonstrates that neural nets can discover features about the world without being told what to look for, and can use those features to solve problems. This ability is an important departure from conventional computers and may enable neural nets to solve problems that ordinary computers can't. Traditional artificial intelligence has produced "expert systems"; one called Prospector, for example, gives advice about where to look for deposits of ore. Another, called Mycin, helps makes medical diagnoses. Expert systems typically have a data base full of information about a subject and a set of rules to make decisions.

Useful expert systems have been difficult to create, however, because they are based on the premise that you can ask experts what rules they follow when they make a decision and what kinds of information are important for that decision. The problem is that experts often don't know exactly what rules they follow. Many experts say that they follow hunches or that they simply have a "feeling" that the answer is the right one.

In the same way that NETalk can be trained to read aloud, it may be possible with the new learning procedure to train a neural net as an expert system. "What we're saying is let an expert give us a set of typical problems and their solutions, what the input and output should be," says Sejnowski. "Then the network can learn what the correct rules are. Our network will *discover* these rules. We don't have to start with high-level concepts; we can just start with these very little units that have a simple learning rule."

Through their learning abilities, neural networks like NETalk may also yield insights into how humans learn. For example, psychologists have long observed that if a person is trying to memorize a list of facts, he will remember them better if those facts are repeated at spaced intervals rather than presented over and over, all at once. In one experiment, Rosenberg and Sejnowski taught NETalk a list of a thousand commonly used words, then tried to teach the machine a new list of nonsense words, using two learning strategies. In one, the repetition of the new word was spaced between words that NETalk had already learned. In the other, the new word was simply repeated over and over. Rosenberg and Sejnowski found that, as in humans, NETalk had better recall of the new words when the new words were spaced.

The Ghost in the Machine

NETalk isn't actually made of chips, bolts, and other hardware. It exists only in the electronic rumblings of the machine it may someday replace, at least for some applications—a serial digital computer. A conventional computer, with its incredible versatility, can simulate any other information-processing machine. Just as a computer can simulate airflow around the wings of a jet, the computer in Sejnowski's office can represent NETalk's neurons and connections, coded in ordinary computer language, and simulate the activity of the connectionist machine.

The computer simulation of NETalk's activities is a slow and laborious process compared to the real thing. "If NETalk were an actual machine," says Sejnowski, "it would be about half the size of my computer and do NETalk's training session a thousand times faster." But Sejnowski isn't rushing to put NETalk into a real machine. "There's no doubt that if connectionism is successful, there will be new machines built that are based on these connectionist architectures. The only question is *which* machines will be built. There are dozens of connectionist architectures out there, and it's by no means clear what the advantages and limitations of the various designs are. We still are in the very early days of exploring. Who knows which will eventually prove to be the most successful? Last year I worked on one type of machine; this year I'm working on NETalk. Next year I may be doing something else. Once you put something into hardware, you don't have that flexibility."

So Sejnowski is content to simulate NETalk on conventional computers for the time being. And he's more than a little surprised at NETalk's abilities. "All along, we knew reading aloud was a difficult problem," Sejnowski says. "In fact, when we started out, even Geoff Hinton, who's a big believer in these network models, thought that reading aloud was beyond the capacity of such a simple network model. He thought we were going to need a million neurons rather than a few hundred. But we went ahead with NETalk anyway, figuring that at least we might be able to get some insights into ways of improving the performance of neural nets. As it turned out, even such a small net of three hundred neurons can capture 95 percent of the regularity—and not only the regularity but the exceptions, too—of English pronunciation. A traditional cognitive scientist would attack this problem by sitting down and writing thousands of rules and micro-rules and observations about English. It would require an enormous amount of intuition and would be a lifetime of work. But here is a simple three-hundred-neuron network capable of extracting many of the very same rules and regularities as well as the exceptions."

Sejnowski flashes another of his proud paternal smiles. "And it does it overnight."

The Apprentices of Wonder

The truth is, the science of Nature has been already too
long made only a work of the brain and the fancy: It is now high
time that it should return to the plainness and soundness of
observations on material and obvious things.
—Robert Hooke, *Micrographia* (1665)

All kinds of intriguing speculations suggest themselves. At
this rate, the question "What is consciousness?" will again
become respectable.
—John Maddox, editor of the British journal *Nature,*
on the impact of connectionism

Whether connectionism will emerge as the dominant model of the
mind remains to be seen. Neural networks and learning procedures
that better reflect the biology of the brain must still be developed—
it's unlikely, for instance, that neurons are modified by back propagation. It is also unknown how current designs for neural networks, which now use a relatively small number of neurons, will
operate when enlarged to contain huge numbers of neurons. There
are the enormous problems of designing networks that can process
symbols sequentially, represent relationships, and focus attention
on a small part of a problem rather than the whole problem at
once.

Uncertainties over these issues have fueled a growing backlash
from cognitive scientists skeptical of the potential of neural nets.
Some researchers dismiss connectionism as a fad, a symptom of a
peculiar fever that seems to afflict mind researchers every twenty
years or so. In the 1920s, that fever was gestalt psychology. Twenty
years later it was cybernetics, followed by perceptrons in the
1960s, and now connectionism. All these theories were supposed
to explain the mind, say these critics, yet the same thing happened
each time: One set of problems was traded for an even more
difficult set.

Other critics see connectionism's emphasis on training networks by associating inputs with outputs as a dangerous backslide toward behaviorism. But the behaviorists' notion of a reflex mind is certainly not part of the connectionist framework. Indeed, the new connectionism had arisen precisely because researchers have been able to construct neural nets that go beyond simple associations to form complex internal representations.

The reawakening of connectionism has even resulted in the republication of Minsky and Papert's twenty-year-old *Perceptrons*. In a new Preface and Epilogue, the authors reassert their doubts about neural nets and chide the new connectionists for overstating their potential. The sudden influx of research money for neural nets, however, has created an atmosphere that will make it more difficult for some connectionists to curb their enthusiasm. "Let's face it, these models are in their infancy," says Rumelhart. "We try to strike a delicate balance between being optimistic and being realistic. Unfortunately, some neural network people in the business world want to sell their product, and there's a fine line between giving a realistic analysis of your machine and advertising it."

In many ways, the new excitement over connectionism's potential to understand the mind and brain is justified. Connectionists have produced theoretical breakthroughs, such as John Hopfield's insight into how neural networks operate, and have developed new, effective methods for training multilayered networks like the back propagation learning rule and the Boltzmann machine. These mathematical formulations can be understood easily by researchers everywhere—and perhaps improved upon. Connectionists are also the beneficiaries of the explosive growth in the very technology that put them out of business twenty years ago. The price of memory storage in a conventional computer is one hundred millionth what it was in 1950. The development of cheap, fast computers with vast memories means that connectionists can now simulate sophisticated neural networks like Sejnowski's NETalk, which would have been impossible to create 20 years earlier.

Neural nets have the potential to bring scientists in many areas of research onto a common ground. In doing so, connectionism may have an impact similar to the way that Darwin's theory of natural selection revolutionized biology, sociology, psychology, and even philosophy, because it provides a common language that any scientist can understand. "Connectionism is going to be a gathering point of the clans," says Gary Lynch. "It may well turn out that our cerebral cortex works something like Terry Sejnowski's neural network. But if it doesn't, connectionism will still be important, because it led to Sejnowski talking to Lynch, and Churchland talking to Hinton, and so on—and out of that, *something* will definitely happen."

The most important impact of the new connectionism may be on our understanding of how we think. The traditional model of the mind is based on the principle that, at its roots, the mind is an engine of logic, and if the mind does not always behave logically, the problem lies in the brain. But this model doesn't account for the way people actually think. "All the experimental evidence points to the fact that people aren't rational," says Rumelhart. "That is a simple fact that the rationalists refuse to accept."

Neural networks match patterns, make generalizations, merge new situations into old experiences, mirror the structures in their environment, and find the best fit among many possibilities. They aren't good at logic, but their sophisticated abilities suggest a new approach to thinking about thinking. "Connectionism is important," says Patricia Churchland, "because it constitutes the beginnings of a genuine, systematic alternative to the 'grand old paradigm.' Connectionist models illustrate what representations might really be, if not sentencelike, and what neurobiological and psychological computation might really be, if not logiclike. They free us from the conviction that the sentence-logic model is inevitable."

Perhaps the most important result of cognitive research over the last three decades, notes Jerry Fodor, has been the realization that brain and mind scientists walked into a game of three-dimensional chess, believing it was tic-tac-toe. Their research at first produced some remarkable achievements in understanding higher mind functions like logic and mathematics, but in the end, observes philosopher Herbert Dreyfus, "the everyday world took its revenge." The classical model has failed to explain our *common sense*—that effortless, fuzzy, and pervasive aspect of our minds that we use every day to get along in the world. This ability is the essential part of our brain's cognitive powers, encompassing our remarkable abilities to gain insights, understand language, and perceive the world around us.

This common-sense thinking ability may be rooted in how the

brain interacts with the community of other brains around it. If so, connectionism may provide a new way to examine what it is to be rational, moral, and intelligent: It may be that what we call common sense is not the making of logical deductions, but the mind's ability to interweave and incorporate its actions into the complex society of minds of which it is a part. "The question that won't go away," says philosopher Hilary Putnam, *"is how much of what we call intelligence presupposes the rest of human nature?"*

Neural networks operate by mirroring the structures in their environment, structures that arise as families, communities, and societies evolve and adapt through writing, government, and culture. Ironically, this cognitive community provides the foundation of each individual's notion of *common sense,* yet it is the interaction of each individual mind with other minds that forms this cognitive community in the first place. Perhaps the mind pulls itself up by its own cognitive bootstraps, forming a society that reflects in some ways the complex society of neurons in each brain.

Such speculations are still in the realm of philosophy, but as John Maddox, editor of the prestigious British scientific journal *Nature,* wrote in an editorial, connectionism has moved these kinds of "coffee-table questions" a little more into the arena of science. Other questions, such as the nature of consciousness, the meaning of dreams, and the origins of culture, may also come to light someday as researchers further refine the new model of the mind.

Whatever the ultimate model of the mind turns out to be, it's likely that the old-fashioned digital computer will remain top banana in the land of computing machines, at least for a while. Computers do some things very well—things our brains do horribly—and that's the beginning of any beautiful friendship. Who cares if a digital computer doesn't actually operate like a brain? Nobody ever complained that a steam shovel doesn't work like a human muscle.

As a model of human cognition, however, the serial digital computer and the formal system whence it arose may go the way of telephone switchboards, steam engines, clocks, wax tablets, and all the other aspirants as "stand-in" for the mind. It is unlikely that neural networks as they appear today will take the computer's place. The mind is probably not one huge neural network but rather a collection of many smaller cognitive units that work together, and a true model of the mind will be a synthesis that contains elements of connectionism and the classical model. The symbolist serial approach and the connectionist approach, so often portrayed as in conflict, may ultimately need each other.

Most brain and mind scientists think that eventually they will unravel the many secrets of our "engine of thought." But they have no illusions about the difficulty of the task. "There's nothing we're aware of that even remotely approaches the brain in complexity," says Lynch. "There is all this wiring, all these hormones rushing in and out of it, and all this other stuff going on. How are you going to figure out how it works? When you take a good look at it, you can't help but say to yourself, 'Hmmm, that's going to be a hard one—a very, *very* hard one. That's going to take some time.' "

Until that time, perhaps we should savor a little amazement that countless researchers will spend countless hours trying to discover what is going on inside the lump of tissue inside our skulls, a lump that is designed primarily to keep us alive but insists on going beyond its job description to write poetry, send people to the moon, fall in love, and build machines to imitate itself. We can at least enjoy a sense of perverse satisfaction that something we all carry around with us—something that in some respects *is* us—is so damned inscrutable. As the philosopher Richard Rorty once said, "The ineffability of the mental serves the same cultural function as the ineffability of the divine—it vaguely suggests that science does not have the last word." Or as the philosopher Christine Skarda put it when she was asked how it is that the great mental marvels of perception, reason, consciousness, and common sense can all be performed by a blob of neurons strung together chugging electricity: "Nobody has any idea, really. It's just *wonder tissue.*"

Our brain is an intimate stranger, so much a part of us and yet far more intricate, extraordinary, and deeply mysterious than anything else in the known universe. After years of scientific study, we have come to realize that we are not the masters of this complex, majestic organ but rather its apprentices. As we learn more about our "wonder tissue," we will also learn from it, discovering along the way the secrets of ourselves.

Selected References

Ackley, D. H., Hinton, G. E., and Sejnowski, T. J.: A Learning Algorithm for Boltzmann Machines. *Cognitive Science,* 9:147-169, 1985.

Anderson, James A.: Cognitive and Psychological Computation with Neural Networks, *IEEE Transactions on Systems, Man, and Cybernetics,* Vol. SMC-13, No. 5, Sept./Oct. 1983.

Barto, Andrew G., and Sutton, Richard S.: Landmark Learning: An Illustration of Associative Search, *Biological Cybernetics,* 42:1-8, 1981.

Chomsky, N.: *Syntactic Structures,* The Hague: Mouton, 1957.

Churchland, Patricia Smith: *Neurophilosophy: Towards a Unified Science of the Mind-Brain,* MIT Press, 1986.

Clark, S.A. et al: Receptive fields in the body-surface map in adult cortex defined by temporarily correlated inputs, *Nature,* 332:444-48, 31 March 1988.

Cooper, Leon N.: A Possible Organization of Animal Memory and Learning, *Nobel Symposium,* 24:252-64, 1973.

Cowan, W. M. et al.: Regressive Events in Neurogenesis, *Science,* 225:1258-65, 12 September 1984.

Crick, F., and Mitchison, G.: The function of dream sleep, *Nature,* 304:111-14, 1983.

Fodor, Jerry A., and Pylyshyn, Zenon W.: Connectionism and Cognitive Architecture: A Critical Analysis, Cognitive Science Memorandum, COGMEM 29, 1987, *Cognition,* 1988 (in press).

Grossberg, S.: How does the brain build a cognitive code? *Psychological Review,* 87:1-51, 1980.

Hebb, D.O.: *The Organization of Behavior,* New York: Wiley, 1949.

Hinton, G.E., and Anderson, J.A.: *Parallel Models of Associative Memory,* Hillsdale, NJ: Erlbaoum, 1981.

Hofer, M.M., and Barde, Y.A.: Brain-derived neurotrophic factor prevents neuronal death *in vivo, Nature,* 331:261-62, 1988.

Hopfield, J.J.: Neural networks and physical systems with emergent collective computational abilities, *Proc. Natl. Acad. Sci.,* 79:2554-58, April 1982.

———. Neurons with graded response have collective computational properties like those of two-state neurons, *Proc. Natl. Acad. Sci.,* 81:3088-92, May 1984.

Hopfield, J.J., Feinstein, D.I., and Palmer, R.G.: "Unlearning" has a stabilizing effect on collective memories, *Nature,* 304:158-59, 14 July 1983.

Hopfield, J.J., and Tank, D.W.: "Neural" Computation of Decisions in Optimization Problems, *Biol. Cybern.,* 52:141-52, 1985.

———. Computing with Neural Circuits: A Model, *Science,* 233:625-33, 8 August 1986.

Hubel, D. H., and Wiesel, T. N.: Receptive Fields, Binocular Interaction and Functional Architecture in the Cat's Visual Cortex, *Journal of Physiology,* 160:106-54, 1959.

Jones, William P., and Hoskins, Josiah: Back-Propagation: A generalized delta-learning rule, *BYTE,* 155-62, October 1987.

Kahneman, D., and Tversky, A. (eds.): *Judgement under Uncertainty: Heuristics and Biases.* New York: Cambridge University Press, 1982.

Kelso, S.R., Ganong, A.H., and Brown, T.H.: Hebbian synapses in hippocampus, *Proc. Natl. Acad. Sci.,* 83:5326-30, July 1986.

Knapp, Andrew G., and Anderson, James A.: Theory of Categorization Based on Distributed Memory Storage, *Journal of Experimental Psychology: Learning, Memory, and Cognition,* 10 4:616-37, 1984.

Kuhn, T.: *The Structure of Scientific Revolutions.* Chicago: The University of Chicago Press, 1970.

Kuperstein, Michael: Neural Model of Adaptive Hand-Eye Coordination for Single Postures, *Science,* 239:1308-09, 11 March 1988.

Lehky, Sidney R., and Sejnowski, Terrence J.: Network Model of shape-from-shading: neural function arise from both receptive and projective fields, *Nature,* 333:452-54, 2 June 1988.

Linsker, Ralph: From basic network principles to neural architecture: Emergence of spatial-opponent cells, *Proc. Natl. Acad. Sci.,* 83:7508-12, October 1986.

————. Self-Organization in a Perceptual Network, *Computer,* March 1988.

Lynch, G., and Baudry, M.: The biochemistry of memory: a new and specific hypothesis, *Science,* 224:1057-63, 1984.

Lynch, G., McGaugh, J.L., and Weinberger, N.M. (eds.): *Neurobiology of Learning and Memory,* New York: Guilford Press, 1984.

Lynch, G. et al.: Cortical Encoding of Memory: Hypotheses derived from analysis and simulation of physiological learning rules in anatomical structures. In *Neural Connections and Mental Computations,* L. Nadel, ed. MIT Press, 1987.

McClelland, J.L., and Rumelhart, D.E., (eds.): *Parallel Distributed Processing: Explorations in the Microstructure of Cognition, Vol. 2, Psychological and Biological Models.* Cambridge, MA: Bradford Books/MIT Press, 1986.

McCulloch, Warren S., and Pitts, Walter H.: A Logical Calculus of the Ideas Immanent in Nervous Activity, *Bulletin of Mathematical Biophysics,* 5:115-33, 1943.

Minsky, Marvin, and Papert, Seymour: *Perceptrons: An Introduction to Computational Geometry,* Cambridge, MA: MIT Press, 1969.

Newell, A., and Simon, H.: *Human Problem Solving.* Englewood Cliffs, NJ: Prentice Hall, 1972.

Orbach, R.: Dynamics of Fractal Networks, *Science,* 231:814-19, 21 February 1986.

Pearson, J.C., Finkel, L.H., and Edelman, G.M.: Plasticity in the Organization of Adult Cerebral Cortical Maps: A Computer Simulation Based on Neuronal Group Selection, *The Journal of Neuroscience,* 7(12):4209-23, December, 1987.

Pinker, Steven, and Prince, Alan: On Language and Connectionism: Analysis of a Parallel Distributed Processing Model of Language Acquisition, Occasional Paper #33. Request for reprints: Pinker, Department of Brain and Cognitive Sciences, MIT Press, Cambridge, MA 02139.

Rakic, Pasko: Specification of Cerebral Cortical Areas, *Science,* 241:170-76, 8 July 1988.

Randal, Brian (ed.): *The Origins of Digital Computers: Selected Papers,* New York: Springer-Verlag, 1982.

Rosch, E.: Natural Categories, *Cognitive Psychology,* 4:328-50, 1973.

Rosch, Eleanor: Linguistic Relativity, in *Human Communication: Theoretical Explorations,* Siverstein, A.L. (ed.), New York: Halsted Press, 1974.

Rosenblatt, Frank: The Perceptron: A Theory of Statistical Separability in Cognitive Systems. Ithaca: Cornell Aeronautical Laboratory Inc., Report no. VG-1196-G-1, 1958.

————. *Principles of Neurodynamics,* New York: Spartan, 1962.

Rosenfield, Israel: Neural Darwinism: A New Approach to Memory and Perception, *The New York Review of Books,* 9, October 1986.

Rumelhart, D.E., and McClelland, J.L. (eds.): *Parallel Distributed Processing: Explorations in the Microstructure of Cognition, Vol. 1, Foundations.* Cambridge, MA: Bradford Books/MIT Press, 1986.

Sejnowski, T.J., Kienker, P.K., and Hinton, G.E.: Learning Symmetry Groups with Hidden Units: Beyond the Perceptron, *Physica,* 22D:260-75, 1986.

Sejnowski, Terrence J., and Rosenberg, Charles R.: Parallel Networks that Learn to Pronounce English Text, *Complex Systems,* 1:145-68, 1987.

Skoyles, John R.: Training the brain using neural-network models, *Nature,* 333:401, 2 June 1988.

Squire, Larry R.: Mechanisms of Memory, *Science,* 232:1612-19, 27 June 1986.

Sternberg, S.: Memory scanning: Mental processes revealed by reaction-time experiments, *American Scientist,* 57:421-57, 1969.

Thompson, Richard F.: The Neurobiology of Learning and Memory, *Science,* 233:941-47. 29 August 1986.

Winograd, T., and Flores, F.: *Understanding Computers and Cognition,* Norwood, NJ: Ablex Publishing, 1988.

Zipser, David, and Anderson, Richard A.: A back-propagation programmed network that simulates response properties of a subset of posterior parietal neurons, *Nature,* 331:679-84, 25 February 1988.

General Reading

Bloom, F. E., and Lazerson, A.: *Brain, Mind, and Behavior,* New York: W.H. Freeman and Co., 1985.

Changeux, Jean-Pierre: *Neuronal Man: The Biology of Mind,* New York: Pantheon Books, 1985.

Gardener, Howard: *The Mind's New Science: A History of the Cognitive Revolution,* New York: Basic Books, 1985.

Hooper, J., and Teresi, D.: *The Three-Pound Universe,* New York: Macmillan Publishing Co., 1986.

Hunt, Morton: *The Universe Within: A New Science Explores the Human Mind,* New York: Simon & Schuster, 1982.

Miller, Jonathan: *States of Mind,* New York: Pantheon Books, 1983.

Ornstein, R., and Thompson, R. F.: *The Amazing Brain,* Boston: Houghton Mifflin Co., 1984.

Index

Grateful acknowledgment is made to the following for granting permission to use material from their work:

To *Scientific American* for permission to use illustrations by Carol Donner from "The Neuron," by Charles Stevens, *Scientific American*, September 1979, pages 56–57.

To Dr. Geoffrey Hinton for permission to use the illustration that appeared in his book, *Connectionist Learning Procedures*, page 15a.

To The MIT Press for permission to reprint the cover art of their publication, *Perceptrons*, by Marvin Minsky and Seymour Papert.

To Dr. John J. Hopfield for permission to use an illustration which originally appeared in his book, *Brain, Computer, and Memory*, and then appeared on page 5 of *Engineering and Science*, September 1982.

To Dr. William P. Jones for permission to base two illustrations on drawings that appeared in his article, "Back-Propagation," in the October 1987 issue of *Byte*, pages 158 and 160.

To Dr. Richard Durrett for permission to use his "complex system" slides, showing the "percolation of oil through sand."

To *Science* magazine for permission to use a figure which appeared in "Mental Rotation of Three-Dimensional Objects," February 1971, Vol. 171, pages 701–703.

To Dr. James McClelland, James A. Anderson, David E. Rumelhart, and George P. Lakoff for permission to print excerpts from interviews conducted while researching this book.